compost toilets

a practical DIY guide

Dave Darby

L I L I

Published in June 2012 by

Low-Impact Living Initiative
Redfield Community,
Winslow, Bucks, MK18 3LZ, UK
+44 (0)1296 714184

lili@lowimpact.org
www.lowimpact.org

ISBN: 978-0-9566751-1-8

Editor: Elaine Koster
Photos: Dave Darby
Except: Fig 4:   User Musphot on Wikimedia Commons
        Fig 7:   Watercourse Systems Ltd
        Fig 10: Department of Physics, University of Florida
        Fig 11: Natsol
        Fig 12: Elemental Solutions
Illustrations: Mike Hammer
Design and Layout: Commercial Campaigns Ltd

Printed in Great Britain by:
Lightning Source UK Ltd, Milton Keynes

contents

illustrations

about the author

Dave Darby is Director of Low-Impact Living Initiative, or LILI, which he started in 2001 and if any green technology in particular was responsible for that, it was compost toilets. In the summer of 1995, whilst living in London, he applied to join a Permaculture design course being run by Simon Pratt at Redfield Community in Buckinghamshire. The course was cancelled due to insufficient bookings but Simon invited him to come for the weekend anyway, as he was building a compost toilet in one of the outbuildings. Dave went along and helped to build the compost loo, was so impressed with Redfield that he applied to join and, eventually, moved there in the summer of 1996.

The compost toilet was finished by then but wasn't being used very much. The reason was simple – it was outside, and there were several flush toilets inside the main house. It was just too much to ask for people to walk downstairs, outside and into a rather dank outbuilding to use the compost loo, when there was a flush loo in their unit or on their landing. And it was even less likely to be used at night or in the winter.

So one of the first things Dave decided to do on becoming a member of Redfield (with the consent of the community) was to build a compost toilet inside the house. He read everything he could on compost toilets and attended a course at the Centre for Alternative Technology in Wales. The two-room, two-chamber toilet he built in the disused ground floor bathrooms at Redfield is the main focus of this book.

Apart from compost toilets, Redfield also had straw-bale buildings, solar hot water, wood stoves, organic gardens and orchards, woodlands, free-range animals, used lime and eco-paints and one member was making biodiesel from waste cooking oil. He decided to find out if other people wanted to learn about these things (as well as about communities themselves), and LILI was born in 2001.

Dave left the community in 2009 to live in the local town, and walks from there to work for LILI every day. He still changes the wheelie bin compost toilet in the stable block centre, where there is accommodation for LILI courses, every year or so. The wheelie bin toilet is also covered in this book.

Dave ran the compost toilets course in LILI's first year, and produced a manual to accompany it; that manual has now evolved into this book.

introduction

If aliens were to land on this planet, and we were to try and explain to them the way we live on earth, amongst all the crazy things that we as a species get up to, perhaps one of the most difficult to explain would be the process of expending vast amounts of time, energy and money producing and delivering pure, precious drinking water to all homes (in the West anyway), only for people to crap in it, thereby destroying two scarce resources at the same time, and causing disposal and pollution problems in the process.

Crazy as this is, of all the topics LILI is involved with this one causes the most mirth. When we attend events, most people look at our information boards and make interested noises about solar hot water, biodiesel or straw-bale building but chuckle when they come to compost loos and say, 'no way', or 'I can just see my mother using one of those when she comes to visit!'

Having said that, there are lots of people who have seen the light. LILI ran a compost toilets' workshop at the 'Big Green Gathering' one year at 9 a.m. on Sunday morning. I trudged over to the yurt where the talk was to be held, rubbing my eyes and not expecting anyone in their right mind to be there, only to find that around 50 people had turned up.

I'd say that a well-designed compost toilet is the pinnacle of sustainable sewage treatment, especially when combined with a waterless urinal (preferably draining onto a straw-bale to produce compost) and a reed bed and pond to cope with liquid wastes and to provide water for a garden.

Crapping into drinking water and flushing it away to be dealt with at a sewage plant is a terrible idea and very expensive. But it fits the typical modern lifestyle very well. People work 50 hours per week in order to earn as much money as possible and have no time to think about alternatives to the wasteful and environmentally-damaging aspects of their lives. As long as it is quick and easy, and you don't have to think about it too much, most people are happy. Those people don't have time for gardening either, with their busy careers, and so have no need for compost anyway. You, of course, are not interested in this type of lifestyle. You want to work part-time so that you have enough time to install and maintain renewable energy systems, produce some of your own food, maybe build your own house using natural materials, spend time with friends and family and, of course, install a compost loo. This will then mean that you don't have to work 50 hours a week to maintain an expensive lifestyle and you will find yourself on a healthier, happier path, rather than the unhealthy, consumerist, environmentally-damaging path promoted by the advertising industry and unthinkingly embarked upon by most people in the West.

This is why I want to help people learn to build and install compost toilets which are comfortable, indoors, odour-free, and generally 'mother-friendly' (i.e. wouldn't scare your mother too much). My mother wouldn't dream of using a compost loo, let alone peak-knocking (see *use and maintenance,* page 101, for details) or emptying one. But then again she wouldn't be reading this either. Some people are up for it and others aren't. The fact that you are reading this probably means that you are.

compost toilets: a practical DIY guide explains the general principles of compost toilets and how they work, their environmental benefits and also includes a description of, and instructions on how to build, a basic one.

I also outline more 'hands-on' ways of dealing with human waste, and although these have the lowest impact of all on the environment, generally, people don't want to know. That is the great thing about modern WCs and sewerage systems – you don't need to get more involved with your waste than just flushing it away and cleaning the toilet.

I would like as many people as possible to install and use compost toilets, and so have made sure using the design described in most detail in this book shouldn't gross-out the uninitiated too much. It is about as low-impact as you can get, and still (maybe) persuade your mother to use.

Fig 1 shows the 'throne' sitting on top of the chamber. Note the bucket for sawdust, steps up to platform, vent from the chamber through the roof and above the gutter-line, instructions on the wall and waterless urinal in the foreground

This design is based on the compost toilets in use at Redfield Community. They work well, are cheap and relatively easy to build and maintain. I can guarantee that if you build and use this type of compost toilet properly you'll end up with crumbly compost that is indistinguishable in texture, smell, and taste (only joking) to the bag compost that you can buy at any garden centre.

fig 1: working compost toilet

It is possible to buy a proprietary compost loo, see *what can I do? – buying a compost toilet* page 43 for details, but there are problems. If you go for a compact and relatively cheap one, they usually use quite a lot of electricity to dry out the liquids from the waste and the general consensus is that they are too flimsy and unreliable. You can go for a big sturdy model, but then the problems involve finding room for it and the high cost involved. So I recommend building your own.

For me there is a deep satisfaction in being integrated into the land and environment in which we live. At Redfield the members are warmed by fires burning logs that grew here, eating organic food that grew here, giving food scraps to chickens which provide eggs; their bedding and waste goes onto the compost heap, compost goes onto the garden and even human waste goes to the soil around fruit trees, which give us delicious fruit and juices.

Not everyone feels the same way, especially about compost toilets but the more people see well-maintained compost toilets in operation, the more they will become accepted and their undoubted benefits will be better understood. Ecotourism is an excellent way to promote compost toilets as well as a whole range of other eco-friendly technologies. That way people can be introduced to them when on holiday and having a good time, and just maybe they will think about incorporating them into their everyday lives when they get home.

I have used the term 'human waste' above as shorthand for faeces and urine. Many writers on sustainable sewerage systems refuse to use the term 'waste' as it is in fact a valuable resource. I acknowledge this, and only use the term because it is useful shorthand and it refers to the fact that they are waste products of human digestion.

If you intend to use this book to build and install your own compost toilet, remember that a badly-maintained or smelly compost toilet can undo the work of 100 good ones, so try to make it as comfortable and attractive as possible and keep it clean and well maintained. This will help to spread the word. Good luck!

what is a compost toilet?

A compost toilet is a dry or waterless toilet, i.e. one that doesn't use water to take waste somewhere else. Water isn't at all helpful when it comes to dealing with human waste – it has to be separated again further down the line. It's only for transport, because we're not too fond of our waste, and want to get rid of it as quickly as possible. But we don't have to get rid of it; we can turn it into healthy, odourless compost, which would otherwise cost good money to buy.

A compost toilet allows natural processes to produce useful compost after a resting period, the length of which depends on the type of toilet. This is an important point. Old-fashioned, hole-in-the-ground pit latrines can't really be called compost toilets, because, generally, no soak was added, see carbon:nitrogen ratio, page 20, and there was no drain so that the pile could decompose aerobically, see decomposition, page 16, to produce healthy, odour-free compost. They usually became smelly and a new pit was dug elsewhere, or the contents were put onto fields without being properly composted. In the BBC TV programme, Castaway, a group of people on a remote Scottish island were using what they described as a compost toilet, but wasn't. They didn't seem to be putting a soak in and part of their procedure was to take off the lid and turn the raw sewage with a spade (for some reason that was left unexplained). The whole thing was misguided and disgusting, and probably turned millions of viewers off compost toilets for life.

There are many different types of compost toilet. You can buy them off the shelf, but they tend to be expensive. Some have just one chamber, some have a heating element, some have sophisticated ways to separate urine, some are not completely dry, but use tiny amounts of water to 'micro-flush' waste to the chamber. You can buy quite sophisticated compost toilets with a maintenance contract in the States.

At the other end of scale, as described in The Humanure Handbook, see resources, page 119, you can crap in a bucket and tip it onto a compost heap to break down with kitchen and garden waste. It works perfectly well, but I want the use of compost toilets to spread, and I don't feel that too many people will want to do that. This book focuses on a compromise between the bucket system and the off-the-shelf system; a basic loo that is indoors and cheap to build yourself. A lot of people think that a compost toilet has to be outside, in its own little outhouse, but I don't think it will get much use if it's outside, especially in winter.

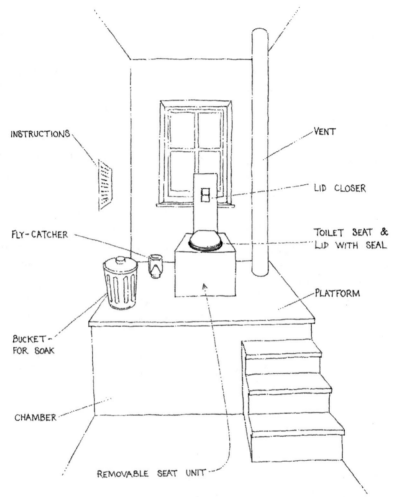

fig 2: components of a basic compost toilet (one chamber shown)

The design described in this book is meant to be installed indoors, although it will work just as well in an outbuilding. There will be some properties where this is impossible but, even then, some sort of extension (maybe straw-bale, as in the Redfield visitor centre) can be built to accommodate it. My design is for a basic toilet that has two chambers, uses no water or electricity, and liquids drain from the bottom via gravity; it is therefore probably the lowest-impact design of all.

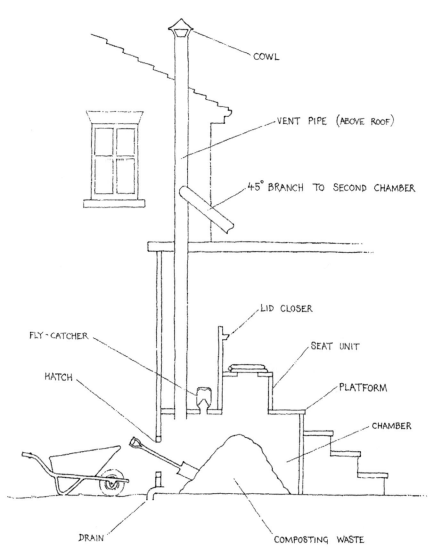

COWL

VENT PIPE (ABOVE ROOF)

45° BRANCH TO SECOND CHAMBER

LID CLOSER

FLY-CATCHER

SEAT UNIT

HATCH

PLATFORM

CHAMBER

DRAIN

COMPOSTING WASTE

fig 3: cross-section of one chamber of a basic compost toilet

components of a basic compost toilet

- two chambers
- platform
- vent
- removable seat
- hatch
- fly-catcher

The basic design has two chambers, but you can buy toilets with one chamber. I think the two-chamber model is better, because it gives the contents of each chamber at least a year to break down with no fresh material added. The new material will contain more nutrients, liquids, salts and ammonia, which means that worms can't be added, and pathogenic material may be leached into the compost you are about to remove. A chamber is used for one year while the contents of the other one are decomposing.

A seat is situated on a platform over the chamber in use. The seat can be removed when the chamber is full and re-situated over the empty chamber; the full chamber is then sealed and allowed to decompose for at least a year. You could also have a seat on each chamber, so that it doesn't have to be removed – but you'd have to make sure that you can't open the lid of the one that isn't in use.

There has to be some sort of hatch to empty the finished compost, a vent to take away odours and moisture, and a fly-catcher is recommended, as it is very unpleasant to open the toilet lid and have flies fly out.

The basic design in this manual has no fans to remove smells and vapours, no electrical elements to heat the waste, no paddles or mixers to turn the pile, or to knock the top off the pile. Some off-the-shelf models have these, and if you want them, by all means get one – they're just more expensive, and our design works!

decomposition

The following is a very basic description of the composting process in the compost toilet. You don't have to worry too much, as nature takes care of it – it works. It is the same process that happens in the wild all the time (thank God – otherwise we'd be wading through fallen leaves, dead animals and poo all the time). Nature has had millions of years of practice and will do the job perfectly.

Decomposition is what happens to organic material on the path to becoming inorganic, or mineral. Everything that was once alive is organic. All our food was once alive, and therefore all our waste is organic.

The work is carried out by micro-organisms, mainly bacteria and fungi. In a compost toilet, or in a compost heap, bacteria tends to work on the fresher material near the surface, and fungi breaks down the older material in the middle of the heap. In our design, it is mainly fungi that do the work, as the heap doesn't get too hot. You may have noticed your compost heap (especially if you have just added a lot of grass cuttings) can get very hot. This is the action of bacteria. But your compost toilet pile won't get this hot; it will usually operate between 15-30°C, at which temperatures the decomposition will mainly be carried out by fungi and moulds, although there will be some bacterial activity.

In organic matter, molecules contain long chains of carbon atoms. When organic matter decomposes the bonds in these chains break and energy is released which the micro-organisms use. Oxygen is used in breaking these bonds and is converted to water. Chains become smaller and smaller each time they break down, until they won't break down or decompose any more. The material can then be considered inorganic, or mineral.

The process is the same as in a compost heap but in a compost toilet, although there may not be enough material to generate much heat, due to the time involved between use and emptying, organic material will all break down to compost. It would probably be more accurate to call this a mouldering process rather than a composting process. However, it produces compost, and the phrase 'mouldering toilet' is, if anything, probably less attractive to most people than compost toilet.

95% of all living tissue is made up of 4 main elements:
- oxygen (O): 62% (mainly in the form of water)
- carbon (C): 20%
- hydrogen (H): 10%
- nitrogen (N): 3%

Fats contain mainly carbon and hydrogen.
Carbohydrates contain mainly carbon, hydrogen and oxygen.

Proteins contain mainly carbon, hydrogen, oxygen and nitrogen. Other important elements present in organic matter that contribute to the formation of nutrients on decomposition are:
- phosphorus (P)
- sulphur (S)

aerobic decomposition
Aerobic decomposition occurs when organic matter decomposes in the presence of oxygen. This is the type of decomposition we're after in the compost toilet.

Aerobic decomposition is around 20 times faster than anaerobic decomposition.

Elements combine with oxygen to produce odourless, harmless and/or useful (to plants) end-products:
- oxygen   -   oxygen (O_2)
- carbon   -   carbon dioxide (CO_2)
- hydrogen   -   water (H_2O)
- nitrogen   -   nitrates (NO_3)
- phosphorus   -   phosphates (PO_4^{3-})
- sulphur   -   sulphates (SO_4^{2-})

Oxygen and hydrogen have many ways of escaping, in combination with other elements.

Carbon is converted to carbon dioxide and leaves as a gas.

Nitrogen, found mainly in proteins, is very important in that it forms nitrates which are the most important nutrients for plants and they, paradoxically, cause the most problems in watercourses via algal blooms.

It eventually ends up either as nitrogen gas, or as part of the proteins of living tissue again after it is taken up as nitrates by plant roots.

Here is a brief outline of the path it takes:
Firstly, it is excreted from the body as urea (there are lots of different versions of the chemical formula of urea: $CO(NH_2)_2$; $(NH_2)2CO$; CH_4N_2O; CON_2H_4 – the figures all add up to the same combination of elements though), which then degrades to ammonia (NH_3). Ammonia then uses oxygen (quite a lot – around 5mg of oxygen per 1mg of ammonia) to create nitrites (NO_2), in combination with a type of bacteria called nitrosomonas.

Then the same process produces nitrates (NO_3), with another type of bacteria called nitrobacter. These processes also require carbon and don't work too well in the cold (less than 5°C).

Then, either:
1. the nitrates are taken up by plant roots, or
2. micro-organisms use some oxygen from nitrates (especially in the absence of free oxygen) to decompose more organic matter, leaving nitrites (NO_2), nitric oxide (NO), nitrous oxide (N_2O), or (by far the most abundantly) nitrogen gas

Phosphorus combines with oxygen via the action of micro-organisms to produce phosphates. These are very useful to plants, but are the main cause of eutrophication (algal blooms) if they get into watercourses.

Sulphur is present in small quantities in all organic matter. As organic matter decomposes it combines with oxygen to produce first sulphites (SO_3^{2-}), which then combine with more oxygen to become sulphates (SO_4^{2-}). If the decomposition is aerobic, they cause no problems.

anaerobic decomposition
Anaerobic decomposition occurs when organic matter decomposes in the *absence* of oxygen.

Elements then combine with hydrogen to produce end-products which can be smelly, potentially explosive and not so useful to plants:
- oxygen   -   water (H_2O)
- carbon   -   methane (CH_4)
- hydrogen   -   hydrogen (H_2)
- nitrogen   -   ammonia (NH_3)
- phosphorus   -   phosphane (PH_3)
- sulphur   -   hydrogen sulphide (H_2S)

This is the type of decomposition that occurs when the micro-organisms can't get any oxygen, i.e. if the heap is waterlogged. We certainly don't want this kind of decomposition in the toilet – ammonia and hydrogen sulphide (the rotten eggs smell of stink bombs) stink to high heaven, and methane and hydrogen are explosive! As for phosphane, it will only be present in small amounts, which is good, as chemical industry health and safety data indicates that it is both explosive and may cause asphyxiation in enclosed, poorly-ventilated areas.

So we absolutely have to ensure that oxygen is present in the heap, and the most important factor in achieving this is to ensure that there is a permanent drain at the lowest point of the chamber, and to check that it never gets blocked.

You shouldn't wee in the compost toilet if you can possibly help it (and let's face it, you often can't); small amounts are OK, but don't use it just for a wee. As well as making the heap waterlogged, urine breaks down to ammonia, which gives off a foul smell and actually kills useful micro-organisms.

Another reason that the heap should be kept aerobic is that certain microbes that feed on bacteria need oxygen to survive. If the heap starts to turn anaerobic these microbes will begin to die off and allow bacteria numbers to spiral out of control; their dead bodies will clog any air gaps that might be left and contribute to the general slimy, smelly nature of an anaerobic pile. Yuk.

carbon:nitrogen ratio

Old-timers often say 'oh yes, we used to have those kind of loos when I was a child, so isn't this a bit of a backward step?' But old-style pit latrines were not compost toilets because they didn't balance carbon and nitrogen.

Put simply, this means that bacteria and other micro-organisms like to eat a balanced diet (of carbon and nitrogen). Human waste contains too much nitrogen and not enough carbon to maintain the right balance. This means that the micro-organisms have to give off some excess nitrogen, and they do this by combining it with hydrogen from water to produce ammonia (NH_3), which is smelly and very undesirable in a compost loo. This will also mean that there will not be as much nitrogen in the finished compost, which is therefore less useful for plants.

The ratio of carbon:nitrogen in the compost toilet should ideally be around 30:1. This is the ratio at which bacteria / micro-organisms like to 'eat' waste.

Below are the ratios of carbon:nitrogen in human waste:
- C:N ratio in faeces   -   8:1
- C:N ratio in urine   -   0.8:1

The first thing to notice is that urine should be kept out of the pile because of the enormous amount of nitrogen compared to carbon. So a permanent drain on the chamber and facilities for people to wee somewhere else are essential.

The next thing to consider when trying to maintain a carbon:nitrogen (C:N) balance is a soak. A soak is a carbon-rich material that is put down the toilet after each use. Examples include:
• **sawdust or wood shavings** provide an excellent soak as they cover immediate smells too; they also have a large surface area to make carbon available to the micro-organisms. The C:N ratio of sawdust is around 500:1, so it is an excellent source of carbon, and only a small handful need be dropped into the toilet after each use. Small wood shavings are probably better than sawdust, especially very fine sawdust, as this won't aerate the pile very much. The carbon in sawdust and wood shavings is in the form of lignin, which breaks down much more slowly than the carbon in, for example, straw and paper, which is in the form of cellulose; this hasn't been a problem in my experience though
• **straw and hay** aerate the pile better than sawdust, but don't cover immediate smells so well. The C:N ratio of straw is 120:1, of hay – 60:1
• **shredded paper** works too, but may contain nasty inks and dyes and doesn't cover immediate smells
• **compost or soil** can be used, but are not as effective as the above materials as they don't contain so much carbon
• **lime or wood ash** have also been used and they are good at covering immediate smells, but don't have enough available carbon to balance the nitrogen; actually ash isn't a good idea either in compost toilets or on compost heaps as it is inert (and so doesn't compost) and contains salts that can kill useful micro-organisms
• an unusual but extremely good soak is **popcorn**. It would work out quite expensive to use it regularly but, if you have any stale popcorn, it will aerate the pile and allow oxygen in for the micro-organisms to do their work

It's up to you what you use, and it probably depends on what's available. If you live on a farm or a smallholding and you chainsaw firewood, or you live near a sawmill, then sawdust is a good idea. If you live on, or close to, a farm that has a lot of straw, then straw will work for you.

In Romania there is a problem with 'dunnies' overhanging rivers and streams, so that human waste causes pollution problems in watercourses; there is also a problem, now that Western companies have moved in to fell their vast forests, with conifer sawdust leaching into watercourses and causing excess acidity. If only they could combine these two problems, not only would they solve them, they would produce a resource in the process.

I've used sawdust and hay, and found sawdust to be the best soak. Sawdust, because of its larger surface area and smaller particles, produces a drier, more compost-like end product, makes more carbon available and decomposes more quickly and completely.

The soak material can be kept next to the toilet in a bucket with a lid and one handful, or a little scoopful, is enough. Toilet paper and the cardboard middle are fine too, although it's probably best to use plain, unbleached paper.

The soak is useful in more ways than just adding carbon:
• it stops the pile becoming anaerobic by aerating it and allowing oxygen in
• it covers immediate smells
• it soaks up excess liquid, and helps stop the pile becoming waterlogged

So, to maintain the carbon:nitrogen ratio, and to stop the pile turning anaerobic, it is essential:
• not to have too much urine in the pile
• not to allow the pile to get too wet, although micro-organisms need moisture to work, but not too much
• to allow air (oxygen) into the pile, so that any excess nitrogen will combine with oxygen to produce nitrates rather than ammonia
• to put carbonaceous material into the toilet
• to have a drain on the chamber

pathogens / hygiene

Ten reasons not to be worried about pathogens from compost toilets:
1   pathogens would have to be in the humans who use the toilet in the first place
2   someone would have to have contact with the waste, which isn't possible inside the chamber
3   if pathogens like the conditions inside the human body (as regards temperature, acidity / alkalinity, moisture content etc.), they certainly won't like it outside the body. Almost all human pathogens will be dead after a few hours outside the host
4   all human pathogens will be dead within two months outside the human body, with the possible exception of two species of roundworm eggs
5   those two roundworm species are tropical and so, to get them, the compost toilet would have to be used by someone who has been in the tropics and was unlucky enough to pick up those particular roundworms

6   the roundworm would then have to survive and lay eggs, which would have to survive one or two years in a compost pile, then manage to get onto food in the garden and still be there after harvesting, washing and cooking

7   if the pile reaches 55°C for three days (easy to achieve in the summer) then absolutely all pathogens will be killed

8   if the waste is being composted then the environment is much more hostile to pathogens than in a septic tank, yet septic tanks are accepted. Tests in the US found that compost from a Clivus Multrum compost loo contained 10,000 times less pathogenic material than sewage sludge from a septic tank. And yet farmers are still allowed to spread sewage sludge on their land. Would you rather eat food produced on land which has had compost from a compost loo, or sewage sludge? Well, you already are eating food from land which has had sewage sludge on it

9   some micro-organisms actually produce antibiotics that kill pathogens during aerobic decomposition

10 the issue is acceptability rather than risk. 3,000 people a year are killed by cars in Britain and around 30,000 either directly or indirectly by alcohol; and yet cars and alcohol are completely accepted in society. The risks from a compost toilet, if properly used, are infinitesimally small compared to the much greater risks that we come into contact with every day

Having said all that, it is vital to deal with any waste water (urine, drain from chamber) properly, because there are some nasty waterborne diseases that can be spread via 'black' water. Please see the *what happens to the urine section?* of the *what can I do*? chapter, page 57, for more detail.

Flies can be a vector for pathogenic material; but any flies that walk on you, your food, or in your kitchen have been walking on something gross in the not-too-distant past anyway. It's what flies do; they are part of nature, humans have always lived with them, and they probably help build people's immune systems. We shouldn't expose ourselves to obviously pathogenic situations, but equally we shouldn't try to hermetically seal ourselves into sterile bubbles either.

People often see compost toilets as a retrograde step back towards the unsanitary conditions before flush toilets were developed. Although the Romans had sophisticated flush toilets and sewer systems, it is true that through the dark ages and up to the twentieth century unhygienic means of dealing with human waste caused smells, diseases (cholera, typhus, dysentery, typhoid fever) and death in most cities and towns. But the point is that the human waste wasn't composted, it was just thrown into the streets, into open sewers and into watercourses.

Rivers through cities became open sewers themselves. Problems occurred because people came into direct contact with pathogens in faeces; flies were able to spread disease; but mainly because the waste came into contact with water. Water spread the pathogens much more effectively and allowed disease into the drinking water supply. Also, carbon-rich material was not added and you didn't end up with useful compost.

In the second half of the nineteenth century in Britain, when sewers and WCs were starting to be introduced, death rates from diseases traceable to faeces actually increased, because it was a new technology and there were lots of teething problems, like leaks from sewers into drinking water.

Also, there was no black death in Asia, where faeces were traditionally not thrown into the street or taken away by water, but by ox-carts to spread onto agricultural fields. There weren't usually problems in rural areas where pit latrines were used either – it was only the combination of faeces and water that caused problems. Having said that, leachate from pit latrines can get into the water supply by seeping into groundwater and into wells and watercourses.

Compost toilets don't allow faeces to get into the water system and they don't allow contact by people or flies with anything until it is no longer recognisable as of faecal origin, but is pathogen-free, useful compost.

And the alternatives are more risky, if anything, for human health. We should be thankful that dumping of sewage sludge at sea is no longer allowed, as sewage then ended up on beaches and caused a direct health hazard if people came into contact with it.

The current system is not much better either; sewage treatment plants use large amounts of chlorine, which ends up in rivers and is considered a cancer risk. Even McDonald's are phasing out chlorine-bleached french-fry bags because of the health risks associated with chlorine. And sewage sludge – pathogens, toxic chemicals, heavy metals and all – is still applied to agricultural land while the leachate, of course, ends up in groundwater and watercourses.

It could get worse too; many sewage plants no longer work properly because industrial toxins kill the useful bacteria in the plant. There have been proposals in the US to use nuclear waste to sterilise sewage. Just how far down the wrong path can you go?

a bit of history

Using human waste as fertilizer has a long tradition in other parts of the world. In the Far East there is a tradition of using faeces for 'night soil' – so called because the waste was collected at night on carts to be spread on agricultural land. This also happened to some degree in Europe, but not on the same scale as in China and Japan. Although the waste wasn't composted at source, it eventually decomposed in the soil. However, it is better to compost the waste at source because, until it is composted properly, nitrogen and other nutrients are water soluble and can leach out of the soil.

Often though, the use of human wastes in the Far East has been romanticised somewhat, and although, as mentioned before, there was no Asian black death, I would like to stress that it is an *extremely* bad idea to use uncomposted human waste on agricultural land or gardens. Although the use of night soil meant that water wasn't wasted or polluted and soil structure wasn't damaged, it also meant that there has been a long history of related problems. Even up to the twentieth century, it has been reported that over half of China's rural population had some kind of tapeworm or fluke, and that deaths from faecal-borne diseases were disproportionately high.

I have used toilets in India that were over a pig pen, and the pigs were waiting for a meal below – this is a very old and a very good way of turning human waste into non-human waste very quickly.

Nowadays, two countries – Mexico and Vietnam – have a distinct leaning towards compost toilets. Apparently there are 100,000 twin-chamber compost toilets in those two countries; they've obviously seen the light.

In the West, in the late nineteenth and early twentieth centuries, we moved away from pit latrines, night soil, or disposal in streets and rivers, to disposal by flushing with water – the WC or water closet. I will come back to this later, but first let me mention a man who campaigned against this waste of water and this turning of two resources into a problem, and promoted the earth closet. His name was Henry Moule, and his earth closet competed very closely with the water closet before eventually succumbing towards the end of the nineteenth century.

Henry Moule was a vicar in Dorset, who invented something that looked very similar to a water closet, in that it had an overhead hopper and a chain. But instead of flushing the toilet with water it deposited a small load of earth. He had recognised some big problems with water closets in particular that they flooded existing cesspits, which created a health hazard and wasted water. When this happened in Henry's case, he filled in his cesspool and got his family to use buckets, the contents of which he buried in trenches in his garden. He found that when he dug in these trenches several weeks later, all he found was soil, with no trace of human waste.

In his closet, the bucket of earth / waste mix was taken out when full and mixed in a pile with fresh earth. He also realised the fertilizer value of this mix, and used it in his garden.

Earth closets became popular in Europe but not in America and, when the sewer systems were completed in most Western cities, the water closet was finally victorious as most people wanted the waste out of sight immediately, whatever the health or environmental consequences.

It was a closer contest than you might think though, and several public institutions changed from a water system to an earth system. In doing so, they saved money and solved lots of problems, such as toilets becoming blocked by people flushing unsuitable things down them.

The 'out of sight, out of mind' attitude prevails today too. Perhaps in future, new homes will include a proprietary compost toilet, which can be emptied and the contents used by the owner / occupier or, for a charge, emptied by local authority trucks and taken to be spread on agricultural land. (This way those who help reduce the number of truck-miles clocked up to remove the waste will save money too.) Then none of us will waste water when we flush the loo, we won't need to spread sewage sludge with industrial toxins onto farm land, and we won't need the expensive and energy-intensive sewerage system. Until then, you can build one or install one yourself.

fig 4: Henry Moule's earth closet, improved version ca. 1875

benefits of compost toilets

comparisons with other systems

I'll briefly describe some other sewerage systems, and then compare the performance of compost toilets with these systems as regards the environment. The systems are:
* conventional system to sewage plant
* septic tank and leachfield
* small rotating-arm sewer
* cesspool
* reed beds
* other types of toilet
* no treatment at all

conventional system to sewage plant
The most common scenario in Britain; most houses are on mains sewerage, so after people flush the toilet that's the last they want to see, hear or know about their waste. Here's what happens to it, very briefly:
* waste pipes take domestic sewage to sewage treatment plants
* first solids such as tampons, condoms, rags, paper, plastics, sand and grit are removed by screens, and disposed of in landfill
* then faeces are allowed to settle in large settling tanks (this is called sewage sludge), so that the liquids can be separated for further treatment
* usually, liquids are passed via rotating arms then through a filter or percolating bed consisting of stones on which micro-organisms that feed on nutrients and other micro-organisms in the liquid waste are growing
* sometimes oxygen is blown through tanks containing the liquid waste, again to encourage beneficial bacteria to consume nutrients and other micro-organisms; this is called aeration
* sewage sludge is then either heated and digested anaerobically by bacteria, to produce soil conditioner which is spread on agricultural land, incinerated or taken to landfill
* often there is further disinfection of the liquid waste by ultra violet light or chlorination before it is allowed into water courses or the sea

septic tank and leachfield
Fewer people in Britain have this system (although millions have it in the US); generally it's for people in rural locations in properties without mains sewerage.

Waste from the house flows first to a tank where solids settle, and liquids are allowed to flow out and into a series of underground pipes with holes in; these are laid in a trench filled with gravel, called a leachfield (or drainfield, or soakaway system). The liquid seeps through the holes and into the gravel, then into the soil where it is purified by soil micro-organisms which take up the nutrients that the liquid contains. The water is then either taken up by plants, falls to the level of groundwater and then, eventually, horizontally into streams, or evaporates.

fig 5: this septic tank (made from concrete, half hidden by vegetation and dead leaves) holds solids and allows liquids to pass through to a rotating-arm sewer; a truck pumps out the contents once a year

In the tank, some of the solids float (scum) and some sink (sludge). There is some decomposition in a septic tank; by anaerobic bacteria on the sludge and by aerobic bacteria on the scum. Periodically a truck will come (Redfield's has 'The Poo Lorry' in big letters on the side), insert a large flexible pipe into the septic tank via a manhole, suck up all the solids and take them away. This costs around £150 a time and, if it's done annually, compares well to normal sewerage charges. Sometimes, for a single household, the truck won't need to come every year – only every two or three.

Some septic tanks have two chambers, and tend to be made of concrete, but newer ones can be made of reinforced fibreglass.

small rotating-arm sewer

Some larger properties can have their own rotating-arm sewer, similar to the large ones found in sewage treatment plants. Bacteria on the stones will remove nutrients and harmful micro-organisms, and the water quality of the outflow will be checked periodically by the Environment Agency. The finished liquid can then be released into water courses.

A septic tank will collect solids before the liquid part of the waste enters the rotating-arm sewer. The mechanism works using gravity so there is no electricity required for pumps, etc. However, some maintenance is required – each month the rotating arms need to be removed and cleaned out, and the stones turned.

Redfield has a septic tank and rotating-arm sewer, see fig 6.

fig 6: rotating-arm sewers use no electricity and are very durable; this one was built by the council for Redfield when it was an old peoples' home in the 1950s, and it's still going strong

cesspool / cesspit

Cesspools or cesspits (there is no difference) are used where no consent has been given for a soakaway system, and the property is not on the mains sewerage system. Usually the reason that you can't have a leachfield or soakaway is that the property has clay soil. Cesspools can be bought off the shelf – they are usually cylindrical and bigger than a septic tank, because all solids and liquids are contained within it – there is no overflow to any kind of drainfield. For this reason they have to be emptied by a tanker much more regularly than a septic tank, although some older ones (which are sometimes built of brick or stone) allow liquids to soak into surrounding soil.

reed beds

Reed beds can be used to purify liquid waste instead of the bed of stones used in rotating-arm sewers or sewage treatment plants. They are artificial mini-wetlands specifically built to purify waste liquids. Reeds are planted that capture oxygen from the air and deliver it to their roots, where it is used by micro-organisms to help consume nutrients and possible pathogens. A reed bed will usually have primary treatment supplied by a settling tank, and then liquids will flow out of it into the two types of reed bed mentioned below.

fig 7: horizontal-flow reed bed

There are two types of reed beds – vertical-flow and horizontal-flow:

1 vertical-flow reed beds provide secondary treament. They don't contain water all the time – water enters at the top, percolates down and leaves at the bottom; they operate aerobically
2 horizontal-flow reed beds provide tertiary treatment, and do contain water – if you imagine a bath filled with gravel, which is filled from one end and overflows at the other; they operate less aerobically and more anaerobically than vertical-flow beds

The sludge that settles in the settling tank can be removed by a tanker as with a typical septic tank, or it can be pumped into a 'sludge bed' planted with a different type of reed. After several years, the sludge will be composted and can be dug out to use in the same way as compost from a compost loo (much harder work though).

Reed beds can be used in conjunction with a compost loo, so then you won't have to dig solids out. That way the reed bed just deals with urine, greywater and the leachate from the compost loo chamber. Even better – you can use nutrients in the urine for fertilizer; see the *what happens to the urine?* section, page 57. I would say that possibly the best system for a dedicated low-impacter is a compost loo, waterless urinal onto a straw-bale, and a reed-bed leading to a pond, next to the garden so you can dip your watering can in. Then frogs and toads (very good for the garden, as they eat pests) can breed in it (introduce spawn from another pond and they'll keep coming back). Greywater can be directed to the reed bed too. It can be a beautiful facility as well as a useful one.

other types of toilet
There are other types of toilets available, usually for people off mains sewerage, or for boats, mobile homes etc. These often involve mixing waste with chemicals (usually formaldehyde-based – nasty), which are then tipped down a conventional toilet or into a sewer drain. The chemicals don't help in any way other than to control smells and, of course, they hinder decomposition in the sewage plant, and cause toxicity problems in watercourses. The last thing we need is the application of more synthetic chemicals in a world already overloaded with synthetics that the environment (including humans) hasn't evolved to cope with. Having said that, you can now get environmentally-friendly additives to stop smells.

It's very difficult to think of a way to use compost toilets on a boat, where space is at a premium. The Envirolet and Biolet, see the *buying a compost toilet* section, page 43, are just about small enough in some cases but use a heating element to evaporate liquids, which is a waste of energy. The best option is probably a chemical toilet using 'green' chemicals, with a 20-litre sealed container underneath which you empty about once a week (into a sewer pipe at an official mooring, into a conventional toilet or, if you are in a remote location, you could bury it).

Other toilets incinerate waste, using large amounts of energy to destroy a resource and cause pollution; and yet others coat waste in oil, further complicating the situation with a pollutant, and causing more problems than they solve.

We're going to discount these 'solutions' as too much of an environmental nightmare to discuss.

no treatment
In less-developed countries there may be no sewage treatment at all, and so sewage is either deposited directly into rivers, in woods or waste ground or in holes in the ground. It was common to see people using the rivers as toilets in rural India when I was there 10 years ago.

Alternatively, there may be flush toilets and what appears to be a conventional sewerage system, but the proper sewage treatment facilities may not be in place and so raw sewage is dumped straight into rivers. This is the case in Romania.

Raw sewage contains pathogens in the form of bacteria, viruses and worms.

reducing water pollution

Micro-organisms are much more efficient at getting oxygen than larger creatures like fish. If all the bacteria and other micro-organisms in raw sewage were pumped directly into rivers, they would use all the available oxygen and fish would die. The main measurements of the health of a river as regards sewage are BOD (biochemical oxygen demand) – the amount of oxygen being demanded by micro-organisms, and SS (suspended solids) – measured using a filter; large numbers of particles kill larger creatures by blocking light.

Algae are particularly problematic as they grow very rapidly, causing 'blooms' which use all the oxygen in the water and also block out the light; ensuring that watercourses become green and slimy and devoid of life (except algae of course).

Phosphates are the major cause of algal blooms (eutrophication) in rivers, by providing nutrients for the algae to feed on. About half the phosphorus in sewage plants comes from human waste, and the other half is from phosphates in washing powders (they soften the water to make soaps more effective).

Nitrates are also a cause of algal blooms. Ironically, they are causing problems (or requiring sewage farms to avoid these problems) when they could be doing a useful job providing nutrients to plants on land. Compost toilets ensure that plants get these nutrients and then they don't have to be dealt with as if they were a nuisance.

Compost toilets score well as regards water pollution as solid waste is dealt with on site and doesn't get anywhere near watercourses and, anyway, when organic matter is composted properly the nutrients are no longer water soluble and so won't leach out of the soil into groundwater. Also, the more compost toilets there are, the less water will be flushed to sewage treatment plants. This is a good idea as the more water that is used, the more chance there is of flushing nutrients through the system and into the rivers.

Depending on what happens to the urine, it could end up in watercourses. Peeing on straw-bales, leachfields and reed-beds are all good options that avoid this.

Conventional sewage treatment prevents most of the nutrients and suspended solids in the sewage from reaching rivers, but not all, and uses an awful lot of chemicals and energy in doing so.

Raw sewage going into watercourses is obviously the worst option. This is exactly what happens in most cases in developing countries though.

water saving

If all the fresh water on earth was put into a cube its dimensions would be 95 miles on each side. If you think about the size of the planet, and the 6 billion people on it, as well as the billions of tonnes of biomass of other species, then that really is a tiny amount. And this amount of fresh water is fixed, whilst the human population, and our *per capita* water use, is growing. Water shortages are already a serious problem in drier parts of the world and will become more so in future, possibly sparking resource wars. As long as less-developed countries are following the Western model of flush toilets (and they are seen as a measure of development for everyone to aspire to), then this problem is going to get a lot worse. As more and more water is extracted from rivers for sewage transport less is available for wildlife and sometimes the flow of the river itself can be threatened.

It's up to all of us to try and use less water and avert these potential problems.

Around 32% of domestic water use is for flushing toilets, and so installing a compost loo can immediately reduce your water use by a third. An average western family of four will flush the toilet say 15 times per day and, at around 12 litres a flush, that's 180 litres a day. Multiply this by 365 to get an annual total of 65,000 litres down the loo each year. Building regulations now stipulate a 7.5 litre flush, and this will be reduced further to 6.5 litres soon.

Low-flush toilets are available which use even less than this but not as little as compost toilets, which don't use any water at all.

With regards to water saving, compost toilets win against all other systems except no treatment (which doesn't use water, but contaminates it), as all the others use water to flush away the waste.

soil improvement

During the twentieth century, almost half the topsoil was lost from the agricultural land of Western countries. You have to take a moment to let that sink in. A century in which half of our most precious resource was eroded due to incredibly unsustainable farming practices. The use of chemical fertilizers, especially, meant that it was no longer necessary to add organic matter to the soil to increase its fertility. It could be done instantly, in a modern, clean way by adding chemicals manufactured in factories and delivered in plastic sacks on trucks.

Unfortunately this meant that no humus was being produced. Humus is the rich dark top layer of the soil, which contains lots of decomposed and decomposing organic matter. Humus provides food for micro-organisms, nutrients for plants, it helps retain water and, very importantly, it provides structure for the soil and prevents erosion.

So, as the amount of humus in the soil fell, the soil became less fertile and so more chemicals were added until we were on a vicious downward spiral of ever-more chemicals and ever-less soil.

The human body has evolved over millions of years to thrive on food grown in rich, healthy humus. It can't change in such a short time to cope with foods grown in an artificial, humus-poor soil. Cancer is on the march everywhere in the developed world and it's hardly surprising when we steer so far from the natural world in which we evolved.

And of course plants won't be as healthy when grown chemically rather than naturally. Humus holds a lot of water, and also nutrients that are not in solution, so plants can take either water or nutrients – whichever they need. Chemical fertilizers are water soluble so when the plants take water they also take nutrients, whether they need them or not. Chemically-grown fruit and vegetables tend to be big and rather tasteless as a result.

A compost toilet of course produces compost and using that reduces the need for chemical fertilizers which destroy soil structure.

It's true that there is (anaerobic) decomposition in a septic tank. At Redfield there are even tiger worms busily eating the scum in the tank. However, having emptied some of the scum from a septic tank, I found that there's not the same level of decomposition as in a compost toilet. Compared to the crumbly compost that we get from the compost loo, what comes out of a septic tank is, well, not very nice. Neither is the anaerobically-digested sludge from sewage farms as good a soil conditioner as compost, even if you ignore the fact that it will contain industrial toxins (see below).

If an individual wanted to be self-sufficient in food the nutrients in that person's urine and faeces, together with the composted waste from the previous crop, would be more than enough to provide all of his or her food.

other benefits

no chemical cleaners or bleaches
You will, of course, only use eco-friendly cleaning products in your compost toilet (for example, to clean the chute), and you would never dream of using chlorine-based bleaches as they would kill the friendly bacteria working away down there. Unfortunately though, most people with flush toilets do just that (and toilet blocks are as bad as liquid bleaches). Chlorine reacts with organic matter to produce 'chlorinated organics' which don't break down in the environment and so accumulate over time. They belong to the same family of chemicals as DDT and poly-chlorinated biphenyls (PCBs) – both banned in the seventies – and are toxic, and possibly carcinogenic (Greenpeace and Breast Cancer Action say they are, the chlorine industry says they're not).

no sewage sludge problem

1.4 million tonnes of sewage sludge are generated in sewage treatment plants each year in Britain. Dumping at sea was banned in Britain in 1998, and now half is spread on agricultural land and the other half is either sent to landfill or incinerated. After it's incinerated, the ash goes to landfill.

Firstly, how many truck-miles and how many litres of fossil fuels are required to transport that lot around? (I don't know the answer to this one, but it's a lot.)

Secondly, in landfill it leaches into the groundwater and gives off greenhouse gases (to be fair, many landfills now collect the methane to burn for electricity generation; our local landfill has 10 miles of pipes inside it collecting the methane. But this isn't happening in poorer countries. A landfill I visited in Romania was not only not collecting gases, it was on fire!).

Thirdly, sewage sludge contains industrial waste and so incineration will emit mercury, lead, cadmium and dioxins, as well as carbon dioxide, into the atmosphere. Flue gas treatment can remove some but not all of this.
So let's think about agricultural use and the industrial waste that is part of the sewage sludge from sewage treatment plants. The waste will include heavy metals and up to 60,000 different industrial chemicals, many of which are toxic and/or carcinogenic, and very few of which have been tested. And they talk about the risk of pathogens from compost toilets!

There was a successful campaign to stop sewage sludge being dumped at sea ('Surfers against Sewage' were very active) in 1998. There are now several websites dedicated to opposing the use of sewage sludge on agricultural land, but I couldn't find one that mentioned compost toilets, which is surprising because it's the only method of sewage disposal, except for reed beds, which doesn't lead to potentially pathogenic and/or toxic-waste-laced sewage ending up in the soil.

Enlightened countries such as Sweden (it's always Scandinavia isn't it?) have banned the use of sewage sludge on agricultural land; but it's difficult to think of an environmentally-friendly way of disposing of millions of tonnes of sewage sludge containing a cocktail of toxic chemicals and heavy metals. It would, of course, be much better not to generate the stuff in the first place and this could end up being the best reason of all for the promotion of compost toilets.

no greenhouse gas emissions

First a quick summary of the 'greenhouse effect': the whole spectrum of solar radiation passes through our atmosphere to warm up the earth. However, reflected radiation that bounces off the earth is long-wave only. There are some gases in the atmosphere that allow solar radiation through but absorb the reflected long-wave radiation, stopping it from escaping into space, and so causing global temperatures to rise (glass does the same thing – hence the greenhouse effect). No climate scientists these days argue that global warming isn't happening and very few think that it isn't down to humans, although those that do get a disproportionate amount of media coverage, and the International Botanical Congress and the UN Environment Programme both state that current warming trends could mean the loss of up to 60% of all species of plants and animals in the next 100 years. Combine this with predicted sea-level rises of tens of metres if current trends continue, and the implications for humans are obvious – drought, famine and more ecological damage that will ultimately threaten our survival.

Methane and carbon dioxide (CO_2) are greenhouse gases and, although CO_2 is the most important because of the large amounts of it in the atmosphere, methane is better at absorbing long-wave radiation.

Compost toilets break down human waste aerobically and, although it does produce CO_2, which will be taken in by the plants that the finished compost helps to grow, this process doesn't produce any methane.

In all the other systems mentioned the waste breaks down anaerobically releasing methane as well as CO_2.

saving energy

Compost toilets not only save water, but they also save all the electricity used to pump it to our houses (along with all the emissions of CO_2, sulphur dioxide, nitrogen oxides, carbon monoxide etc. associated with electricity generation). They also save all the energy, hardware, chlorine and other chemicals used to extract and purify drinking water.

All other systems, except no treatment, which really isn't a sensible option, use water to flush away the waste.

Compost toilets also save all the energy (and associated emissions) required to deal with sewage, including all the electricity used in sewage treatment plants, and all the fuel in the trucks that empty septic tanks and cesspools and transport sewage sludge.

very low overall resource use
You could spend a whole day thinking of all the resources used in conventional sewerage systems – not only water, electricity and chemicals but trucks to move sewage sludge around, pipes to deliver water and remove sewage, heavy equipment used to lay these pipes, factories to manufacture this equipment – and so on. Suffice it to say that compost toilets don't need any of these things.

saving money
* no truck to remove solids from septic tank – about £150 a year
* no Environment Agency discharge licence (if you collect urine, or have a reed bed / leachfield system) – about £900 a year
* reduced water charges: more and more people are on water meters and all new dwellings have them. The average charge is between £1-2 per cubic metre. Annual family savings on flushes works out at around 65,000 litres. There are 1,000 litres in a cubic metre, so that's 65 cubic metres, saving between £65-130 a year
* low-maintenance, less need for plumbers (who can be very expensive)

what can I do?

very basic ways of dealing with human waste

The simplest way to deal with your waste is to compost it directly. Have a bucket with a toilet seat over it, throw in a handful of soak (e.g. sawdust or straw) when you use it and, when it's full, tip it onto a heap, and add straw, hay, garden waste and kitchen waste.

The advantage of this system is that it is very cheap and easy, with all the benefits of compost toilets – you keep all your nutrients and don't use water to flush them away.

The main disadvantage is that you have to handle it before it's composted, when you are transporting it in the bucket from the toilet to the heap. Most people would not like to do this, although it works well and, if you're up for it, there's no reason why not. Well, except that you'd probably have to live quite remotely as neighbours wouldn't like it, it's labour intensive and uncomposted human waste can contain pathogens, and these can be transported back to humans via flies. Also, you wouldn't want kiddies to play anywhere near the heap, in case they came into contact with pathogens. Apart from that, it's fine! For more details try *The Humanure Handbook,* see *resources* page 119.

buying a compost toilet

There are many different types of compost toilet that you can buy, they come in many different sizes, shapes and styles, and with a range of different prices.

Here is a description of several types of proprietary compost toilet. We haven't included enormous, expensive ones intended for public buildings or models unavailable in Britain. The details of British distributors are listed in *resources*, page 113.
- Aquatron: www.aquatron.se, Swedish company. Works with a low-flush toilet, liquids and solids are separated by the momentum of the flush; waste falls into a composter, and liquids go to a UV unit which kills pathogens. Price is around £450 plus VAT for the separator which sits above the chamber. You can build your own chamber and attach the separator to it. Distributed in Britain by Elemental Solutions

fig 8: the Aquatron separator

- Biolet: www.biolet.com, US company. Small, electric heater, mixer (electric or manual). Price: c. £700. distributed in Britain by Wendage Pollution Control

fig 9: basic Biolet toilet

- Clivus Multrum: www.clivus.com, Swedish. Probably the most famous compost toilet company; it has a large single chamber that is usually sited in a cellar. The chamber has a sloping floor with a series of baffles so that materials can take years to get to the bottom where there is a hatch for emptying. There are several different models and sizes. They work very well but are a bit expensive for the domestic scale but are good for public facilities. Price: around £3500 for a domestic installation. Distributed in Britain by Kingsley Clivus

fig 10: composting chamber of a Clivus Multrum

- Compus II: this is a twin-vault compost toilet manufactured in the UK by Natsol Ltd. – www.natsol.co.uk. Price: around £2350 installed.

fig 11: Compus II

• Dowmus: Australian company, the website was www.dowmus.com but the company has gone bust so it is now defunct. This is a large, single chamber model which takes kitchen waste and greywater as well. Solids are removed via a chute with a screw mechanism, and liquids go off to a leachfield. It has been distributed in Britain by Elemental Solutions. They sell various parts of the kit, such as pedestal, drainage kit, low-wattage waterproof fan etc. The price depends on how much of the kit you buy, and how much you decide to build yourself. There is a forum for Dowmus owners who are having problems with their unit – http://dowmus.hinternet.com.au

fig 12: Dowmus pedestal

- Envirolet: www.envirolet.com, manufactured in the US; will ship to Britain. Similar to Biolet, electric and non-electric versions, small, plastic casing, relatively cheap. Price: from £1200

fig 13: basic Envirolet toilet

- Rota-loo: Australian company, www.rotaloo.com, has a rotating chamber with several different compartments which rotate when full, so that an empty one is under the seat, while the full ones are decomposing. Price (not installed): around £1500 for a family-sized system. Distributed in Britain by Maurice Moore

fig 14: different-sized Rotaloos and compartments

- Soltran: a module designed to house a compost toilet (usually a rota-loo), with a solar collector to speed up composting and evaporate urine. Australian design, distributed in Britain by Maurice Moore
- Sun-mar: US company, www.sun-mar.com, heating element, hand crank for rotating a drum to mix and aerate the waste. They produce a range of different models, including the Ecolet toilet for boats. Price: c. £1000 for smaller models. Distributed in Britain by Eastwood Services
- Biobag: very simple and cheap system based on a basic seat and a bag. See www.kernowrat.co.uk. Also the Separell system

There are case studies of Biolet, Clivus and Dowmus in *Lifting the Lid*, see *resources* page 119.

I think that maybe the small plastic toilets are a bit too flimsy, too small, and have too much that can go wrong. Compost toilets need to be sturdy and very reliable. A bad experience with something like toilets can put people (especially slightly squeamish people) off for life and they'll never be persuaded to try compost loos again. Also, small proprietary toilets often have lots of fiddly things to do as regards moving chambers, or taking chambers out to empty them. This is a major drawback, as most people understandably want as little to do with uncomposted waste as possible. Toilets therefore should be as low-maintenance and reliable as possible.

Toilets with heating elements are not really composting, or even mouldering, toilets anyway as waste is desiccated not decomposed. It could become re-hydrated when added to soil resulting in smells and potential pathogens due to insufficient decomposition. And the heating elements use significant amount of electricity (which is why using electricity for heating is almost always a bad idea).

Micro-flush toilets that use a very small amount of water won't hurt the composting process, as long as there is a drain at the bottom. But the process works without water, and you'll save more water if you don't have a flush – micro or not, so why bother? You can also buy vacuum toilets that suck excrement away to a holding tank; they can be found on aircraft. They don't use water but they do use electricity to suck a vacuum. Again, why bother? The only reason could be that you just don't have room for a chamber under your toilet, or you have room, but no way of emptying it. A vacuum toilet could remove waste to anywhere you want it.

To reiterate, I think that designs have to be solid, reliable, with no chance of mishaps – so perhaps it's not a bad idea to dispense with ideas of urine separation, electrical devices or vacuums that can fail, or fiddly things to empty relatively regularly. It only takes one bad experience with something as unpleasant as faeces to put people off for life.

building your own compost toilet

This, of course, is my favoured option and I have provided a full description of how to install the two-chambered type of compost toilets that were built in the main house of Redfield Community. Please see the *step-by-step DIY guide* chapter that starts on page 63 for details.

other types of home-built compost toilets

wheelie bin loo

In addition to the two-chambered compost toilet in the main house at Redfield there is a different type in the converted stable block visitor centre – and it is based on using wheelie bins.

fig 15: the seat above the wheelie-bin loo, with bucket of sawdust, and instructions on the wall

A straw-bale extension was built at the back of the centre and, as there is a drop in the ground level at the front and back of the centre, we were able to support the extension on 4"x2" timbers so that there is a drop of around 1.5m (5ft) to the ground, which a wheelie bin could fit into.

Inside the extension a washbasin and a toilet seat were fitted over a chute made from a 25-litre plastic bucket with the bottom cut off. The wheelie bin fits underneath this, and is sealed all round with thick black plastic sheet and gaffer tape. The toilet is used for a year and then the gaffer tape is removed, the bin is wheeled out and the lid is put on (with some bricks on top so that no-one is tempted to open it too soon). This bin is allowed to rest for a year while its contents decompose, and another wheelie bin is wheeled into place to replace it. This way you don't need two rooms / two seats – you continue to use the same seat, and alternate the chambers.

fig 16: the Redfield stable block wheelie-bin compost loo underneath the straw-bale extension (clad with tongue-and-groove)

After the first wheelie bin has rested for a year it can be emptied, its contents used for fertilising fruit trees etc., and then swapped with the second, and so on.

Some adjustments were needed to this system though:
1 because there is a chute, which is smaller than a box on a platform, it can pick up deposits on its sides – so it needs to be cleaned occasionally, using eco-cleaners and a toilet brush
2 the chute was made from white plastic buckets. This needs to be painted with black bituminous paint – white is not a good colour for a compost toilet chute, for obvious reasons. Black camouflages things a lot better

fig 17: the previous year's wheelie bin 'resting' in an out-of-the-way place to allow the waste to decompose

3 there has to be a drain at the bottom of any chamber, and the original drain on the wheelie bins was a tap, which we drained into a container around once a month or so and then emptied away into a sewer drain or in the woods. This liquid surprisingly didn't smell too badly – it was a bit like the 'worm tea' that drains from a worm bin (or a bit like 'baby bio' or compost liquid fertilizer with a slightly earthy smell). So not as unpleasant a job as you would think; however, having a tap allowed the very bottom of the pile to turn anaerobic, meaning that even after a year the bottom wasn't composted properly. So now a permanent drain has been installed (a copper pipe falling gently to a sewer drain, see fig 18) and it's working much better

fig 18: the drain pipe at the bottom of the wheelie bin slopes gently to a sewer drain.

4 the wheelie bins don't have a vent. I thought this might mean they would begin to smell – especially in summer. This hasn't been the case though, but should they, a piece of drainpipe could be installed on the side of the building, with a cowl above the roof line, a hole could be cut into the side of the wheelie bin, near the top, and the end of the pipe inserted through the hole (there would need to be an elbow on the pipe to take it into the wheelie bin). Silicon sealant would then be applied around the hole.

fig 19: the fly-trap for the wheelie bin, consisting of a plastic bottle containing some water

fig 20: the chute into the wheelie bin from the room above is secured with gaffer tape to keep out flies

temporary oil drum loo

There was a mini-festival at Redfield for a weekend a few years ago. There were around 150 people camping and so we set up a little temporary compost toilet in the sheep-shed, based on an oil drum. We fixed a seat on top, installed a vent but no drain on the bottom. After the event, we simply removed the oil drum to an out-of-the-way place to allow its contents to decompose anaerobically for three years (anaerobic decomposition is slower than aerobic). The top has to be covered so that no water gets in while it's resting.

Alternatively we could have installed some sort of drain, or even a tap from which liquids could have been drained into the sewer or the woods after the event, and then the contents could have decomposed aerobically.

fig 21: temporary 'oil-drum' compost toilet, with vent pipe and straw-bale steps

straw-bale compost loo

'Green and Away', an outdoor environmental conference centre in Gloucestershire, has lots of environmental facilities, including a type of compost loo I'd never heard of before I went there. It comprises a portable cabin on stilts over a chamber made of straw bales. It's used over the course of several conferences or festivals (in the normal way by throwing in a handful of soak – usually sawdust – after use), then the cabin is removed, and the straw bales are pushed in to cover the pile. The floor of the chamber is the earth, which easily absorbs any excess liquid, and the pile is left to compost for one or two years. Grass grows over it and it's pretty much hidden, and not at all an eyesore. The portable toilet room can then be positioned in a new location over another straw bale chamber. Apparently local environmental health inspectors visited and were happy with it.

This system is obviously only suitable for rural locations, and only then if you're sure that children won't be able to play on the pile.

compost toilet extension or outbuilding
You could have a room housing a compost toilet as an extension to an existing building. However, if the building is of natural materials such as straw bales, timber, rammed earth etc., then you have to remember that the walls of the chamber will still have to be brick / concrete block / plastic etc., and the floor concrete. You can then clad the chamber in natural materials.

fig 22: another temporary compost loo, this time with biodegradable walls!

what happens to the urine?

The first question is: do you wee in the compost toilet chamber or not? If it's just a wee you want, then the answer is simple: no. Too much liquid in the chamber will cause the system to become anaerobic, resulting in some pretty nasty smells. Also, urine contains salts that many organisms don't like (worms for example) and if it turns to ammonia they like it even less. Get people to wee somewhere else, and this somewhere else will depend on your location, your climate and your outlook on life.

Urine can be a useful thing to collect as it actually contains more nutrients than faeces. It contains much more nitrogen than faeces (up to 90% of the nitrogen we excrete is in urine), three times the potassium and up to twice the amount of phosphorus. And no pathogens (not in developed countries anyway, although in developing countries it could contain things like liver flukes) – it is sterile, so there are no health worries, just smells. In fact, it's probably fair to say that if you really want to use human waste as fertilizer the best thing to concentrate on is urine.

As there is so much nitrogen in urine bacteria have to give off lots of excess in the form of ammonia, which means that you can't really store it for more than a day – it becomes smelly very quickly.

Urine could replace chemical fertilizers in feeding the world's population (although solid organic matter would have to be added as well to provide structure), instead of being treated as a problem (which it is, if we're trying to stop it ending up in watercourses).

urine separation if you just want a wee
- you could just go outside in the bushes if you live in a very rural location but, unless you live in a warm climate, this is going to be problematic in the winter.
- you could install a separate urinal. This could be waterless, and could even be a unisex version. You can build your own, or you can buy off-the-shelf waterless urinals too, see *resources*, page 113. If looked after properly, waterless urinals actually smell less than flush urinals because calcium in the water in flush urinals will react with uric acid in the urine to produce more limescale than with water alone. The limescale then absorbs more urine, which encourages bacterial growth and can cause smells. See the a *step-by-step DIY guide*, page 63, for how to build and install waterless urinals. See below for options as to where the drain from urinals should go.

- pee on a straw-bale to eventually produce compost – urine doesn't contain pathogens, but lots of nitrogen. Straw contains lots of carbon, and so the combination is a perfect recipe for producing good compost. You can put the bale in any out-of-the-way place. It won't smell because the bacteria will have a 'balanced diet' of carbon and nitrogen and so won't have to give off ammonia (their way of removing excess nitrogen).
- collect urine in a container with a funnel, and tip it onto your compost heap when it's full. This will benefit the heap, especially if it has a lot of carbon-rich material like dry leaves, twigs, straw or hay. Alternatively you could tip the container onto a straw-bale (see above). You could even have a urinal above the container. This may be better than having to go outside for a wee, especially in the winter, and especially for women, who have to expose a bit more of themselves to the cold. It's safe to irrigate your garden with urine if you want to as urine is sterile, but dilute it with 10 parts water for each part urine as its salt content will be too high for most plants to take and they will die.

urine separation when using the compost loo
If it's just a wee you want of course, it's easy to separate. Otherwise it's not so easy. I've seen compost loos with notices asking users not to pee in the chamber, but let's face it – sometimes you just can't help it. So if you accept that you can't avoid it, then these are the possible approaches to take:
- ensure that there is a drain. I recommend that there should definitely be a permanent drain on the chamber, even if you use one of the separation methods below. It's vital that no-one comes into contact with the liquids from this drain though. It's not like 'worm tea' – the liquid that drains from a worm bin, even though it's the same dark brown colour. Worm tea won't contain pathogens, but liquid from a compost toilet might.
- however, if you don't have a separate provision for urine, a drain alone may not be enough to stop the heap becoming anaerobic and giving off smells. You could install some sort of stainless steel or plastic funnel or trough inside the seat (at the front) to separate urine from solids. The drain from the funnel or trough will take away the liquids in the same way as for urinals (above). The potential problem with this system is that toilet paper can be accidentally dropped into the funnel or trough, then you can end up with soggy blobs of toilet paper blocking the drain. The chances of this increase if a small child uses the toilet or you could even end up with faeces in it, which would be a bit of a nightmare.
- a way around the trough or funnel clogging problem is to build yourself a 'curvy' trough from stainless steel. In this case, the liquids will hit the stainless sheet, and run down it, around the curve and into the trough;

paper or faeces dropped onto it will fall into the chamber and miss the trough. This seems like a lot of trouble to go to, when you could use one of the 'out-of-toilet' urine separation methods.

where does the liquid go from the urinal / urine separation / drain from chamber?

- into a conventional soil pipe or sewer drain. If you install a compost loo where there was a conventional loo before, then there will be an existing soil pipe that the drain can feed into.
- into a sewer drain – underground to the inspection chamber of your sewer drain. You will need to do some detective work to find your sewer drain. Some houses have sewer drains and also surface water drains to take water from the roof and discharge it untreated into a watercourse. However, some houses only have sewer drains (or combination drainage, as it's called) into which all water drains, including rainwater from the roof. This helps to flush the system through but could potentially cause problems by overloading the sewer system during storms. It's up to you to trace your pipes or drains and work out your situation as regards drainage. You have to make sure that you don't connect your urine or chamber waste pipes to a surface water drain. As a last resort, you can use a sewer drain gully outside. This may be OK in a rural location but it is not so good in suburbia, and even in the countryside you'll need a cage round it to stop kids playing with it. This drain should then go to a sewage treatment plant, a septic tank with leachfield, a cesspool, a reed bed or anything else you decide to install.
- to a leachfield; see the *benefits of compost toilets* chapter *comparisons with other systems* section, page 31. You will need consent to discharge and a percolation test for your land. Have a look at the Environment Agency's sheet PPG04 for more information, see *resources*, page 113. N.B. most existing leachfields are too deep, which means that soil microbes can't deal with the nutrients, pathogens or toxins, so the waste water flows beneath the soil and into groundwater effectively untreated. A leachfield must be less than 60cm (2ft) underground for soil treatment to occur. A leachfield consists of a series of gravel-filled trenches containing 50mm pipe with holes in, through which the waste water leaks into the surrounding soil. Micro-organisms in the soil remove nutrients and pathogens – it's the same principle as in a rotating arm sewage bed, where microbes on stones do the work. Soil is an excellent filter for waste water; no better filter has ever been devised. The ideal soil is a loam containing various sizes of particles – not too fine (as this could become waterlogged) and not too coarse (as this drains too quickly and doesn't

have time to treat waste water properly). This is the ideal soil for growing things too. Look out for tree roots near a leachfield, especially willow, which will damage the pipes to get at the water. Also watch out for soil compaction – don't park cars or heavy machinery on the leachfield area. If you already have a leachfield you will be putting less pressure on it by installing a compost toilet and so it will be less likely to fail.

- into a septic tank, cesspool or soakaway (or any other system that exists for waste water – see the *benefits of compost toilets* chapter *comparisons with other systems* section, page 31.
- to a reed bed or pond system – this is above ground so you can see what's happening, there are no underground pipes to get crushed, clogged or damaged by tree roots. It's also prettier.
- how about having a waterless urinal, with a waste pipe that goes through the outside wall and onto a straw-bale?; you can house the straw-bale in a cage or some sort of container to stop kids playing with it. It will compost away nicely, with no pathogen problems. After a year or so, you can shovel out the compost and replace it with a fresh bale.
- even better – have the drain from the waterless urinal go into a pit, or an oil-drum which is half buried, with a hole in the bottom and sitting on about half a metre of small stones, or pea shingle (a soakaway in other words). Fill the pit with straw or sawdust and empty compost from it each year. This will work well on a domestic scale but may become overloaded in a public building.

regulations

First, don't think that there are lots of regulatory bodies out there determined to stop you having compost toilets. All contact I've had with the planning authorities, Building Control, Environmental Health and the Environment Agency has been positive. They just want to be sure that anything you get up to is not going to be damaging to the environment, or cause a nuisance to your neighbours. It was difficult to find anyone who had the first idea about what a compost toilet was but, once I explained, I came across nothing but support.

The Planning Department of my local authority said they have no issues at all with compost toilets, only if there is a new building or extension to house it. They advised us to talk to Building Control.

Environmental Health didn't have any issues either, and didn't know anything about compost loos. They said that they were happy as long as it didn't cause a nuisance to anyone else, for example unpleasant smells. They said that they wouldn't need to come and visit unless anyone complained but thought someone from Building Control might.

Building Control had never heard of compost toilets and there is no mention of them in the Building Regulations. Reed beds, septic tanks, cesspools etc. are covered, but not compost toilets. They suggested that it would be a good idea to submit a Building Control application and then they would talk to Environmental Health. They can't stop you deciding which toilet system to use, providing you don't cause a nuisance to anyone, do anything potentially damaging to your own health or the health of others, or to the environment. If your compost toilet is built and used properly, as described in this book, then no problems should occur and it should benefit the environment rather than damage it.

The Water Supply (Water Fittings) Regulations 1999 do not apply unless there is a connection to the mains supply, which there isn't with a compost loo.

The Environment Agency actively promotes compost toilets (although you'll still have a job finding anyone who knows what they are). They have a range of fact-sheets on ways to save water including one on compost loos, see *resources*, page 113. They also produce several 'Pollution Prevention

Guidance' notes; PPG04 is *Disopsal of Sewage Where No Mains Drainage is Available*, see *resources*, page 113 and covers septic tanks, leachfields and soakaways, and touches on reed beds and compost toilets, both of which are mentioned in a very favourable light. Agency consent may be needed for all of the above, even if discharge is to a leachfield or a reed bed or pond system.

You may need a percolation test to install a leachfield or soakaway and these notes explain how to do it. Contact your local Environment Agency (contact details in PPG04) for more information.

a step-by-step DIY guide

Remember that when building or installing a compost toilet you will need more space than a conventional toilet because waste is stored not simply removed.

tools
These are the tools that you will definitely need:
- mortar board
- shovel
- trowel
- bucket
- lump hammer
- bolster
- SDS drill
- long masonry bit
- power or hand saw
- tape measure
- pencil
- drill and bits
- paint brush
- adjustable spanner
- hammer
- screwdrivers
- masonry drill bits
- silicon sealant gun

In addition, these are the tools that you might need if taking the vent pipe through a flat roof:
- core drill
- core cutter

materials and costs
You could save a lot of money by using second-hand materials. You may have stuff lying around yourself or you could try salvage yards, see *resources*, page 116. If not you will need to use local builders and plumbers merchants or DIY stores. A list of materials and approximate prices for a two-chamber toilet follows.

Materials

item	price	no.	cost £
lime putty (without delivery)	£12.00/tub	1	12.00
sand	£2.50/bag	4	10.00
bricks	40p	c. 200	40.00
SDS drill and long bit (hire / day)	£25	1	25.00
pre-stressed concrete lintel	£20	2	40.00
timber for hatch, seat etc			10.00
coach screws	£1	16	16.00
expanding bolts	£1	16	16.00
screws - 2" no. 8	£6/box	1	6.00
rawl plugs		50	1.00
polypipe plus bends, solvent cement etc			20.00
4x2 timber	£1.20/m	20m	24.00
bituminous paint	£20	1 tin	20.00
3/4" ply - 8'x4' sheet	£28	1	28.00
4" vent pipe	£20/6m	12m	40.00
collars, connectors & brackets for 4" vent pipe			25.00
silicon sealant	£2.00	1	2.00
sealant gun	£5	1	5.00
toilet seat		1	8.00
draughtproof strip		1	2.00
L brackets	50p	8	4.00
cork tiles	£9/m²	4m²	36.00
polyurethane varnish		1 tin	8.00
bucket with lid		1	5.00
Total			**383.00**

Also, you made need to hire a core drill and core cutter – around £25/day.

This is only a rough guide. The costs will vary depending on the size of your project and the materials you use. The total costs above assume that you don't have any of these things, and that you will be buying them from builders' merchants. Obviously it will be much cheaper if you do have them, or if you can find them second-hand.

building your compost toilet

planning

The first step is to sit down and plan the toilet carefully, and think where it will be. These are the basic things to think about:

- will it be indoors or outdoors?
- where will the waste be collected?
- where will the waste be emptied?
- where will the vent pipe go?
- will everything fit?
- is there room for two chambers?
- where will the drain from the chamber / urinal / handbasins go?
- where will the handbasins / urinals go?
- where will the hot and cold water for the handbasins come from?
- will disabled people use the toilet?

Measure your rooms and make diagrams to ensure that everything will fit, including washbasins and urinals with drains. Remember that you will need room for two chambers (although they could both be in one room). An outside wall through which you could put a hatch is ideal. If the room was a toilet before then there will be a soil pipe / sewer drain, which is even more ideal.

Is there room for the waste to drop? It could drop from the ground floor to the cellar (the compost would then have to be carried out in buckets but this may not be too exhausting as it is surprising how the volume reduces as it decomposes), or from the first floor to the ground floor. Alternatively the seat and chamber could be in the same room, but the ceiling will have to be high enough to accommodate steps up to a platform where someone can stand without bumping their head. This is true whether the toilet is in the house or in an outbuilding.

An outside location is fine for your compost toilet if it's only going to be used in the summer – summer use is better than nothing. But if you want to get off mains sewerage, or you want to use it in the winter, then it probably needs to be inside, unless you are very hardy; having said that, outside loos were the norm until the twentieth century. However, the residents of self-build eco-community Lammas in Pembrokeshire (www.lammas.org.uk) having jumped through several planning hoops to be able to start building homes on their community, have fallen foul of building regulations because some of them have built outside compost toilets. Apparently you're not allowed to have an outside loo with a new build nowadays, whatever kind of loo it is. Of course if you live on a smallholding or farm you'll almost certainly never be discovered, but the planners and building inspectors were watching Lammas like hawks because of the ground-breaking nature of their project.

The toilet could be in a purpose-built outbuilding or, better still, an extension and then you can design it so that there is plenty of ceiling height, or the seat overhangs a natural drop to a chamber that can be emptied at a lower level.

If you think that any of the tasks described are too much for you, you could get a builder / plumber in to do them, or you could even consider coming on LILI's *DIY for beginners* course, see *resources* page 113.

A good rule of thumb when cutting pipes, timber, metal or anything else is 'measure twice, cut once'. It can't hurt to measure something twice just to make sure, but it can be disastrous if you cut something without measuring properly first.

The following illustrations, figs 23 and 24, show different ideas for potential compost toilets, indoors and outdoors. Always bear in mind that you have to make sure that the waste can drop somewhere, and be emptied easily.

N.B. not all of these scenarios will work for the disabled.

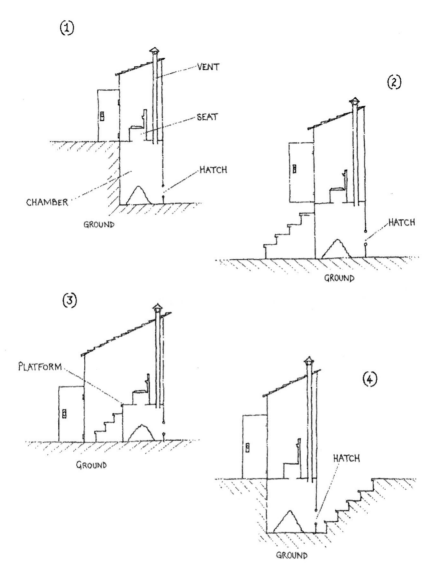

fig 23: various plans for locating compost toilets in an outbuilding;
in no. 4, the steps will need some sort of cover or they will flood

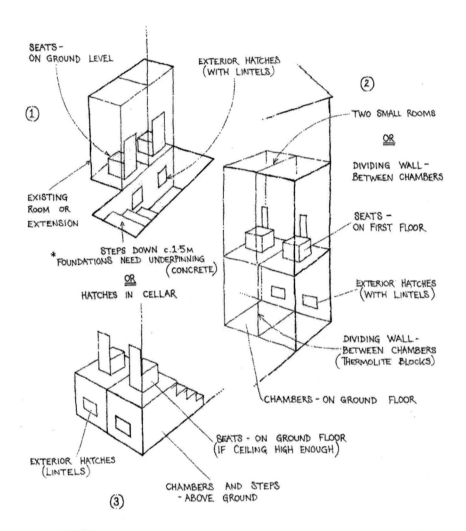

SEATS –
ON GROUND LEVEL

EXTERIOR HATCHES
(WITH LINTELS)

②

①

TWO SMALL ROOMS

OR

DIVIDING WALL –
BETWEEN CHAMBERS

EXISTING
ROOM OR
EXTENSION

SEATS –
ON FIRST FLOOR

STEPS DOWN c.1·5M
* FOUNDATIONS NEED UNDERPINNING
(CONCRETE)

OR

HATCHES IN CELLAR

EXTERIOR HATCHES
(WITH LINTELS)

DIVIDING WALL –
BETWEEN CHAMBERS
(THERMOLITE BLOCKS)

CHAMBERS – ON GROUND FLOOR

EXTERIOR HATCHES
(LINTELS)

SEATS – ON GROUND FLOOR
(IF CEILING HIGH ENOUGH)

CHAMBERS AND STEPS
- ABOVE GROUND

③

fig 24: various plans for locating compost toilets indoors; in no. 1, the steps will need a cover to prevent flooding. A better option might be to empty the chambers in the cellar

The instructions that follow are based on a toilet in one or two rooms on the ground floor but can be amended to fit the various situations illustrated.

construct chambers
See fig 26, page 71.
If the toilet is in constant use, then allow half a cubic metre per person for the chamber. If there are no alternative toilets, then it's definitely better to

err on the large side, but you will be surprised at the small volume of finished compost that you eventually get from the hatch – it will dry out and decompose to around 10% of its original volume.

One person produces around 200 litres of faeces per year; add the same amount of sawdust, but remember that it will decompose to around 10% of volume, and then half a cubic metre per person seems more than enough (half a cubic metre is 500 litres) – but it's better to be on the safe side. Also remember that you won't use the toilet at home all the time (for example, when you're out or on holiday), but this will maybe be balanced by the number of guests who use it.

The larger your chamber, the longer you can let the material decompose, which is a good idea. If you have space build the chamber with one cubic metre capacity per person and use each chamber for two years. Then you'll be absolutely sure that all materials will be decomposed and the final product will be even more like garden centre bag compost.

If the toilet room has a concrete floor, I'd recommend a brick chamber (second-hand bricks of course) because it is stronger than timber when it comes to supporting the platform and, although everything will be painted with a bituminous paint, timber will rot if the paint is breached. You could use concrete blocks instead but concrete is environmentally unfriendly – over 10% of global CO_2 emissions are from the cement industry. If your chamber will be on a timber floor, then it will have to be made of timber – you just have to make sure that it has several coats of bituminous paint so that no moisture can breach it and rot the wood.

Alternatively you could be creative when designing your chamber and possibly make it out of some large plastic bulk container such as the type of IBC (intermediate bulk container) that orange juice is sometimes delivered in – just as long as you can cut into it to make a hatch for emptying the chamber and inserting a drain.

fig 25: preparing a chamber

Fig 25 shows a chamber with steps, platform with cork tiles, vent, and fly-trap. The bricks have been limewashed. The chamber has a 10cm layer of sawdust and is ready for the seat to be attached

I also recommend using lime putty rather than cement when laying bricks, as lime is carbon-neutral; unlike cement it takes CO_2 from the air when it sets. You can purchase tubs of lime putty directly from LILI.

To make a mortar, mix one part lime putty with four parts sharp sand (it's easier to mix putty and sand in a bucket, with a spade, rather than on a board). Soak the bricks well in a bucket of water first as dry bricks will suck all the moisture out of the lime mortar.

If you are building new walls on all sides of the chamber, as opposed to using existing walls for the sides, it is a good idea to leave gaps on the top row of bricks to accommodate cross-timbers which will support the platform, see page 71. Whether the chamber is built of brick or timber coat it completely in bituminous paint, such as the paint designed to seal underneath cars or waterproof guttering and roofs. Alternatively you can do what we did at Redfield and apply bitumen melted in a little wood burner. This will give a thicker coat and definitely won't allow moisture through, but it gives off some hellish fumes. We used an old brush to paint it on then threw the brush away afterwards.

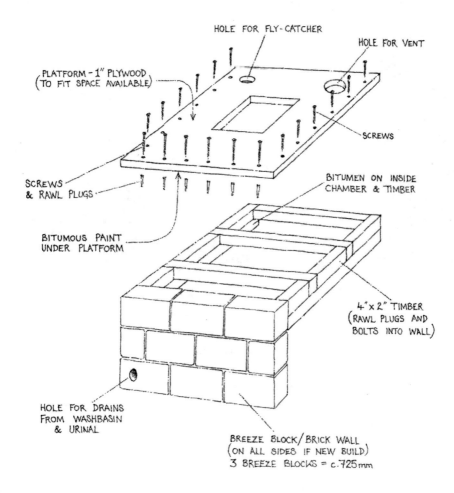

fig 26: plan for the construction of a compost toilet chamber

fig 27: timber supports for the platform completely coated in bitumen

hatch
Please see fig 29 on page 74. Each chamber requires an emptying hatch.

If the wall you're putting the hatch through is brick, then you'll need to mark it, put a few holes around the perimeter with an SDS drill with masonry bit, then knock out the rest with a lump hammer and bolster. The hatch doesn't have to be enormous – just enough to get a long-handled shovel through to empty the chamber.

You will have to insert a pre-stressed concrete lintel above the hole for the hatch, otherwise the bricks above may collapse when they are no longer supported from below.

Then cut some timber to provide a frame for the hatch. Use timber the same width as the thickness of the wall, and then narrower but thicker pieces inside to fix the hatch to, see fig 29. Mortar any gaps between the frame and the wall, then coat with bituminous paint on the inside surfaces only as previously described.

fig 28: access hatch – large enough to admit a long-handled shovel

Then measure and cut some timber for the removable hatch itself. Drill holes around the outside for 8 coach screws (screws with a hexagon head to be used with a spanner) then mark the frame and pre-drill to take the coach screws. Paint the inside surface of the hatch with bituminous paint, allow it to dry then put a line of draughtproof strip around the outside to seal it. Paint the frame and hatch (with eco-paints, naturally), allow to dry then fix the hatch to the frame with the coach screws.

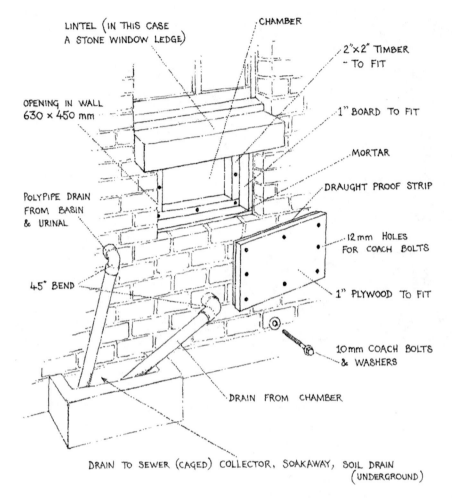

LINTEL (IN THIS CASE A STONE WINDOW LEDGE)

CHAMBER

2"×2" TIMBER - TO FIT

OPENING IN WALL 630 × 450 mm

1" BOARD TO FIT

MORTAR

POLYPIPE DRAIN FROM BASIN & URINAL

DRAUGHT PROOF STRIP

12mm HOLES FOR COACH BOLTS

1" PLYWOOD TO FIT

45° BEND

10mm COACH BOLTS & WASHERS

DRAIN FROM CHAMBER

DRAIN TO SEWER (CAGED) COLLECTOR, SOAKAWAY, SOIL DRAIN (UNDERGROUND)

fig 29: plan for the construction of the hatch for one of the chambers

Some books say that you need a grille or zinc gauze on the hatch, or an air brick somewhere on the chamber (leading to the outside). But, I've found that there are several problems with this:

- icy draughts from the toilet in winter – not very comfortable
- it cools the pile down, and slows decomposition
- it allows light in and means that you can see what's going on in the chamber – not a very good idea
- a grille maybe, and an air brick definitely, will allow flies into the chamber; no-one wants to see flies escaping when they open the toilet lid

install handbasin and urinal

At this point you install your handbasins and urinals because their waste pipes will probably go through the chamber and outside wall to a drain, and it will be impossible to install them once the platform is on the chamber.

I recommend waterless urinals, for obvious water-saving reasons. It's possible to turn any urinal into a waterless one with a Hepworth valve as described below.

You may be able to obtain second-hand washbasins and urinals from salvage yards, see resources, page 113. At Redfield we were given a broken urinal for free, and fixed it with car body filler. See fig 30.

Carefully measure the height of basins and urinals, mark holes for bracket screws, drill holes with a masonry drill, insert rawl plugs and screw brackets to the wall. Basins usually just sit on brackets, which fit into slots, and the basin is held in place by the plumbing and silicon sealant around them – as long as you don't jump on them or deliberately try to pull them off the wall, they'll stay in position. Urinals usually sit on brackets, but also have a fixing at the top which slots over screws to keep them in place. If the urinal is new it will come with instructions and if it's second-hand you'll have to work it out.

When installing the basins and urinals, it's very important to ensure there's enough drop to get the waste water down to the drains. In the case of a waterless urinal this drop also has to incorporate the Hepworth valve.

A Hepworth valve, sometimes called a HepVO (Hep-vee-oh) valve, contains a rubber flap which allows water through but then closes and doesn't allow smells back. The trap or U-bend under the basin performs this function, but you can't have any kind of trap under a urinal as the urine in it will start to smell. Hepworth valves can be bought for less than £10 from any good plumbers' merchant. They have a thread that is compatible with standard waste kits.

fig 30: urinal with Hepworth valve

You can also buy waterless urinals with a little oil trap – urine passes through the oil, which then forms a seal to stop smells. The oil needs to be topped up once a year. I have never come across one of these urinals so can't comment from experience – but the HepVO method is cheap and works well.

Plumb in the hot and cold water to the washbasins. Here isn't the place to describe plumbing techniques. Get a plumber to do it if you are unsure.

Don't forget the earth bonding on any exposed pipework. Again, this is not the place to explain plumbing techniques, so read up or ask a plumber to do it. Basically though, it's a requirement of the Wiring Regulations, and a very good idea, to connect all metal pipework to each other, and then to the earth in your main consumer unit; otherwise, if there is a short-circuit or fault anywhere in your house that renders some metal pipework live, then touching that pipe (or a tap etc.) somewhere else in the building could give you a nasty shock and even be fatal.

female or unisex urinal

Although you can buy a new or second-hand male urinal easily enough, it's difficult to come across a female version (but not impossible, try Googling it). If you're not planning to have conventional toilets as well as a compost loo, then you're going to need a female urinal. The one we installed at Redfield is simply a washbasin fixed onto a welded metal frame with a toilet seat on top, see fig 31.

Choose a deep washbasin (to avoid splashes), or a bidet would almost definitely be better (we didn't have one), and get someone (female, naturally) to use it to see if it works before fixing it permanently. The urinal is unisex of course, so if you only fit one urinal, this is the type to install.

Fit a Hepworth valve underneath the female or unisex urinal in the same way as for a standard urinal, but it is even more important to check that there is sufficient drop to the drain as the urinal will be lower to begin with.

fig 31: unisex urinal

Fig 31 shows a unisex urinal consisting of a seat and lid attached to a washbasin fixed to a home-made metal frame (although a bidet would be better). It has a normal waste kit with a Hepworth valve attached underneath.

drains from handbasin and urinal
The waste pipes, bends, tees and other fittings from the basins and urinals will be 32mm plastic, but be careful because nominally 32mm pipes etc. can be different sizes. Take an example with you when you go to buy the piping. The waste from the urinal can tee into the waste from the basin. As mentioned above, there needs to be a trap under the basin and a Hepworth valve under the urinal. Cut all pipes and fit them together with bends, tees etc. before solvent welding, to make sure that everything fits properly first.

Be sure that there's enough height even after the Hepworth valve for the waste pipes to fall constantly all the way to the drain, so that gravity can remove the liquids. It doesn't have to be a very steep gradient, but it does have to fall a little bit.

Because the chamber is against an outside wall, the waste pipes will probably have to go through it to get outside to the drain (or leachfield etc., see *what happens to the urine?*, page 57. Make holes in the chamber and outside wall with an SDS drill and masonry bit, fit all the pipes and fittings together, then take them apart again, clean them and solvent weld them together. Mortar around the pipes where they go through the chamber and wall (inside and out), and, when dry, paint the mortar with bituminous paint, to stop damp from penetrating into the walls.

drain from chamber
You need a permanent drain from the bottom of the chamber, to ensure that liquid doesn't build up and turn the pile anaerobic.

Make a hole from the inside of the chamber at floor level, through the outside wall, and then insert a length of plastic pipe, with a bend on the outside wall to take the liquids to a drain, leachfield, reed bed etc., see *what happens to the urine?*, page 57.

If you are connecting the drains from the chamber and the urinal to a sewer drain, then the pipe needs to go underground to the inspection chamber of your sewer drain. This isn't difficult – builders do it all the time, and it keeps Environmental Health happy. Make sure it's not a surface water drain though. You need to do a bit of research to find out where your drains go, see *what happens to the urine?*, page 57. If this is impossible to do, then it will also work if the pipes drain into a gully on the surface. You'd need to put some sort of a cage around it to stop kids touching the outlets, and it wouldn't keep Environmental Health anywhere near as happy; but if it's out of the way, it will work.

Mortar around the pipe inside and out and, when the mortar is dry, apply bituminous paint inside. You can put a grille over the end of the pipe inside the chamber to stop it becoming blocked with sawdust, and, to be absolutely sure that no flies can enter the chamber, you can put some sort of mesh over the end of the pipe at the drain – a piece of nylon net curtain fixed with electrical tape will do.

fig 32: here you can see the hatches to each chamber of a two-chamber compost loo

Fig 32 shows the chamber hatches. Each chamber is in a separate room next to each other in the house. The hatch on the left looks as though it has a mesh to allow air in, and in fact it did, but it was covered over when we realised how cold it was to use the toilet in winter! You can see four drain pipes coming through the wall. The lower two drain the bottom of the chambers, and the higher two are from the urinals / washbasin. All four empty into a sewer drain, although this system is not ideal.

If the chamber replaces a conventional toilet, it might be a good idea for the chamber to drain into the old soil pipe – but put some sort of a grille over it to stop it becoming blocked by sawdust.

steps

If your chamber is directly under the seat, you will need steps up to the platform. At Redfield we found that a brick structure with wooden treads works well – see figs 33 and 34. Alternatively you could make wooden steps, or use part of an old stepladder. We calculated the optimum dimensions for people to be able to step up to the platform comfortably. The bricks were laid with lime mortar and painted with a lime wash.

fig 33: these steps are made from bricks with recycled timber treads

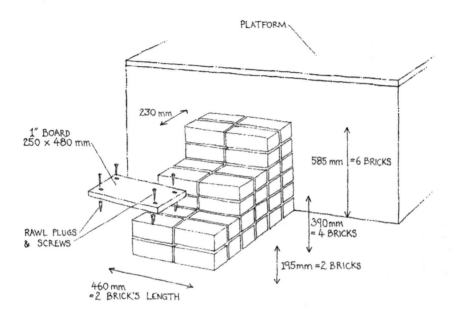

PLATFORM

230 mm

1" BOARD
250 x 480 mm

585 mm = 6 BRICKS

RAWL PLUGS
& SCREWS

390mm
= 4 BRICKS

195mm = 2 BRICKS

460 mm
= 2 BRICK'S LENGTH

fig 34: plan for the construction of the steps up to the platform of one of the chambers

supports for platform

You need some way to support the platform on top of the chamber. A good way is to use 4"x2" timbers (four by twos), which can easily be bought second-hand from salvage yards, see *resources*, page 113. If the chamber is brick (or some other material) on all sides, then the timbers can sit on top of this. It's a good idea when laying the bricks to leave gaps to accommodate these timbers.

If existing walls are used for the sides of the chamber, then you will have to fix some 4x2 to the walls to support the cross-timbers. Measure and mark the correct height, pre-drill the 4x2, mark and drill the wall, and fit expansion bolts (these are bolts with special housings that expand when tightened and pull the timbers tightly to the wall). You then sit the cross-timbers onto the timbers you fixed to the wall, and you can fix more 4x2 between the cross-pieces to secure them in the right place, see fig 26, page 71.

You must remember to then apply bituminous paint to all the timbers – any exposed wood inside the chamber will rot very quickly.

sawdust in chamber

Now is the time to put a layer of sawdust in the chamber – around 10cm (4in) will do. If you want your finished compost to be organic make sure the sawdust comes from the sawing of natural timber, and not from plywood or sterling board, or from treated timber. Wood-preserving treatment involves highly toxic substances such as arsenic, and plywood etc. contains formaldeyde and other substances that you don't want to add to your soil.

platform

The platform can be made from 18mm (¾") ply, cut to size; cut a hole for the seat, bitumen the underside, and screw to bricks (using a masonry drill and rawl plugs) and / or to supports. Apply silicon sealant around the edges of the platform before placing it onto the supports and screwing it into place.

You can lay some sort of floor covering on top of the platform. Cork tiles are good – they are attractive, hard-wearing and natural.

vent

A vent is needed to take away any smells from the chamber. Some people will tell you that you need a fan as well, but I have found that, as long as the lid is kept shut, smells will vent away naturally without a fan.

10cm (4") vent pipe works well and is available from any plumbers' merchant, with associated fixtures and fittings.

Do the difficult part first: take the pipe through the roof or wall, before marking the platform and taking it down to the chamber. If it is a flat, concrete roof, you can hire a core drill (with safety clutch) and core cutter from a tool hire centre (explain to them that the hole is to accommodate a 10cm pipe).

Drill a pilot hole up from underneath first to locate the hole then make the hole using the core cutter from above. Insert the pipe in through the hole, then make it watertight with a collar which is bolted to the roof and sealed with mastic.

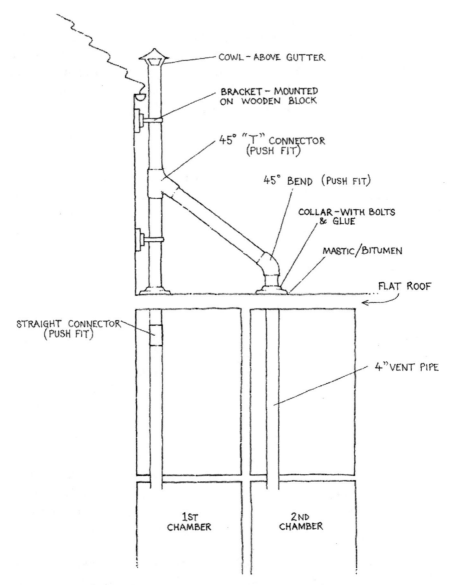

COWL – ABOVE GUTTER

BRACKET – MOUNTED
ON WOODEN BLOCK

45° "T" CONNECTOR
(PUSH FIT)

45° BEND (PUSH FIT)

COLLAR – WITH BOLTS
& GLUE

MASTIC/BITUMEN

FLAT ROOF

STRAIGHT CONNECTOR
(PUSH FIT)

4" VENT PIPE

1ST
CHAMBER

2ND
CHAMBER

fig 35: plan for the vent pipes rising from both chambers, through the roof and above the gutter-line

If the pipe is to go through a slate or tile roof, then a core cutter is not required; remove the slates or tiles, insert the pipe, with collar, and replace the slates or tiles. The slates or tiles should overlap the collar at the top and the collar should overlap the slates or tiles at the bottom. Extra flashing may be required too – there should be instructions with the collar, or ask your plumbers' merchant.

fig 36: to prevent leaks a tight-fitting collar is pushed over the vent pipe after it has passed through the roof

If the pipe is to come through a wall, put a 45° bend inside and outside, so that there is a rise in the pipe at all times (no horizontal lengths). Make a hole in the wall with an SDS drill and bolster as for the drains, insert the pipe and seal around it with mortar.

The pipes from both chambers can come together using a 45° tee into the main pipe that rises above the gutter-line. Do a diagram and talk to your plumbers' merchant, who will be able to give you the right fittings. Bracket the pipe to the outside wall, and take it up above the gutter-line (you don't want smells to be vented into upstairs bedrooms). On top of the pipe, fix a cowl to keep out the rain, and again, a piece of net curtain fixed with electrical tape will keep out flies (although spiders will make webs in the vent pipe to catch flies anyway).

fig 37: the vent pipe is closer to the wall inside than it is outside because of internal plaster. Vent pipes come with special brackets and, to make up the extra distance between the pipe and the wall, the bracket is fixed to a block of wood

fig 38: The vent pipe from the left-hand chamber rises vertically from the chamber, through the roof and above the gutter line. The vent from the right-hand chamber comes through the roof and meets a 45° bend, then rises at 45° until it meets the other vent pipe at a 45° tee. These fittings can be bought at good plumbers' merchants

fig 39: at the top of the vent pipe is a cowl to keep out the rain and insects

Attach a straight connector to the pipe where it comes in through the ceiling or wall, and then drop a plumb line from the centre of the pipe to the platform, and mark the centre of the hole for the pipe into the chamber. With a pair of compasses set to the right radius, mark the hole in the chamber, then drill a few holes around the circumference (with the drill-hole inside the circle) and then cut out the hole with a jig-saw. Cut the pipe to length, push it a short way through the hole in the platform, and then push it up firmly into the straight connector (it should be push-fit); seal around the pipe where it meets the platform with silicon sealant.

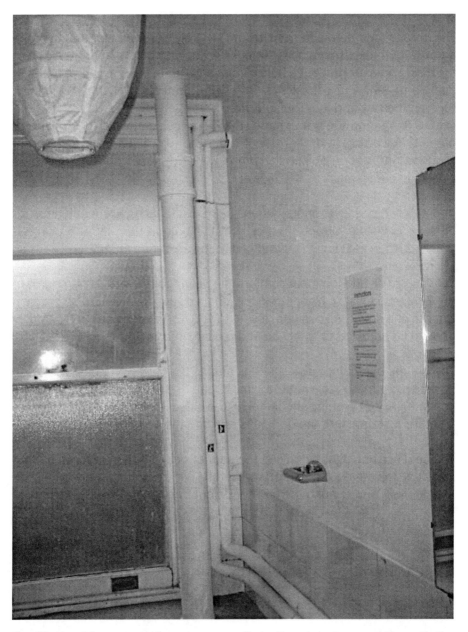

fig 40: the 10cm vent pipe comes up from the chamber and through the roof; in fact the roof work was done first, and a short length of pipe was inserted down through the roof

seat

The 'throne' can be made from 18mm (¾") ply, screwed together to make a box, with a top. Then place a wooden toilet seat on top, mark the size of the inside hole and cut out a hole a little larger than the one you've marked, using a jig-saw. Put silicon sealant on the underside of the seat, screw through the box and into the seat from below, ensuring obviously that the screws don't protrude through the surface of the seat, which could be very painful. Put a ring of draughtproof strip (the type used to draught proof windows and doors) on the lid, which makes a seal with the seat when closed, to keep smells in and flies out.

Apply a generous layer of bituminous paint inside the box, ensuring that no bare ply is visible, attach the box over the hole in the platform using 'L' brackets, then seal it with a bead of silicon sealant all round.

You have to decide the most comfortable height for the box. Most people in the west tend to like it higher than the ideal position for 'evacuation' – which is more of a squatting position. It's up to you. The dimensions in fig 41 may need to be adjusted to what you find most comfortable.

The bituminous paint inside the box means that you won't be able to see any stains, but the fact that the dimensions of the box are larger than the seat means that nothing should touch the sides on the way down anyway – it should just drop into the darkness and into the chamber.

You could save height by having the seat directly on the platform, so that you sit on the edge to use it. You would then almost definitely need to 'peak knock' more and to push the pile back from the front wall of the chamber.

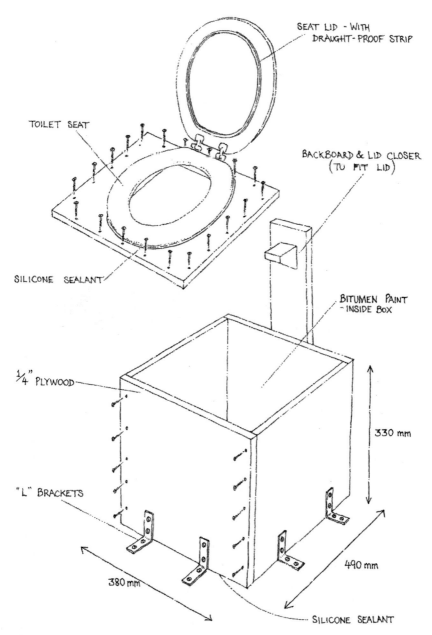

SEAT LID - WITH
DRAUGHT-PROOF STRIP

TOILET SEAT

BACKBOARD & LID CLOSER
(TU FIT LID)

SILICONE SEALANT

BITUMEN PAINT
- INSIDE BOX

¼" PLYWOOD

330 mm

"L" BRACKETS

490 mm

380 mm

SILICONE SEALANT

fig 41: plan for the seat

fig 42: the removable 'throne' over the chamber in use. When this chamber is full, the seat can be removed via the screws in the 'L' brackets around the base and it can be re-situated over the empty chamber

lid closer

When the toilet is not in use you will need to keep the lid closed at all times to keep flies out and smells in. This is very important. If you and your family or housemates are the only people who will use the toilet you may be able to train them to close the lid after use. It may be a bit more difficult with kids, but you have to stress the importance of being absolutely sure to close the lid after every use. You can put a sign on the wall to remind them too.

fig 43: here you can see a ring of draught-excluding strip around the lid to stop smells venting into the room and to stop flies entering

The compost loos at Redfield are used by lots of people, including visitors, and even though we put signs on the wall and inside the toilet door for people to see when they were leaving, and even on the inside of the lid, it was still left open regularly. So we thought of a way to ensure that the lid can't be left open. We put a back-board on the box, with a piece of wood fixed to it that met the toilet lid at a point where the lid could only ever be opened around 89.9° – in other words almost vertical but not quite. This way when someone has finished using the toilet the lid falls back down again.

fig 44: the lid-closer is positioned behind the toilet lid, allowing it to open to not-quite-vertical so that it can't be left open

The only problem with this is one of comfort – it means that the lid is touching your back when you use the toilet. This may be fine if you're wearing work clothes, but not if you aren't. A way round this is to screw two 'eyes' into the side of the lid and the back board, and have a removable hook kept on a ledge above the door – or high enough for kids not to be able to reach it. This is assuming that it's the kids who leave the lid open of course (the kids conveniently get blamed for everything!).

I've been trying to think of some 'Heath Robinson' method of ensuring the lid gets closed, involving pulleys and attachments to door handles etc., so that it's impossible to leave the toilet without closing the lid, unless you climb through the window, or even a lid closer involving a spring on some kind of timer device, but so far I've been unsuccessful.

Self- or soft-closing toilet lids and hinges are readily available on the internet but I haven't tried one yet. Seems like a good idea though.

plumb in handbasin
This isn't really the place to talk about basic plumbing, so if you don't think you can do it, get a professional to do it for you. But if you are confident with basic plumbing here's a brief outline for plumbing-in the handbasins (no water to the waterless urinal of course).

You will need:
- 15mm copper pipe
- 2 isolating valves
- 2 end-feed or compression tees
- various elbows and/or compression fittings depending on which you prefer
- tape measure
- pipe benders
- pipe cutters
- masonry drill
- rawl plugs
- solder
- blowtorch
- flux
- wire wool
- pipe clips
- screws
- screwdriver
- adjustable spanner and wrench if using compression fittings
- 15mm lagging

Turn off the hot and cold water at the relevant stop taps, and empty pipes by opening existing taps. Tee from the hot and cold water pipes, and bring pipework into the toilet. If the room wasn't a toilet before, you may need to knock a hole in the wall to bring in the pipes. Install isolating valves on the pipes once they are in the room. You can then close the isolating valves and turn the water back on elsewhere.

Measure, cut, and fit together all runs of pipe, from the isolating valves to the tap fittings under the handbasins. Put pipe clips onto the pipes and mark where they will go on the walls. When everything fits perfectly, take everything apart. Fit the pipe clips to the walls using a masonry drill / rawl plugs. Then solder all lengths of pipe, elbows and fittings (or tighten compression joints), tighten the fittings under the taps on the handbasins and push the pipes into the pipe clips.

Measure, cut and fit 15mm lagging to the pipes.

fly-trap
A good idea is to install a simple fly-trap. First find a medium-sized jar with a screw-on lid. Turn the jar upside down and place it on the platform in a corner, where you want the fly-trap to be. Draw around the lid with a pencil. Draw another circle about 1cm in from this one, using a pair of compasses. Drill some holes around the circumference of this smaller circle and cut it out with a jig-saw or, if you don't have one, a hacksaw blade.

Cut a similar-sized hole out of the lid of the jar, then apply contact adhesive around the hole in the platform (within the original circle you drew around the lid and on the lid itself. Wait for the contact adhesive to become touch-dry and stick the lid over the hole.

Meanwhile, cut out a piece of perspex to make a little cone that will fit into the jar lid; you can staple the cone together, and / or use contact adhesive. Put the cone into the lid and screw the jar onto the lid as well.

Any flies in the chamber will be attracted to the light, fly through the hole in the platform, out of the small hole in the top of the cone and be trapped in the jar.

fig 45: you can put a fly-catcher like this one in a corner of the platform

You can occasionally unscrew the jar, and vacuum up the flies – or you can sweep them up and tip them back into the chamber.

fig 46: to remove dead flies, unscrew the jar and sweep them up; you can see the perspex cone which allows flies to enter the cone but not get back into the chamber.

blank plate

The seat will be moveable and attached to the platform with 'L' brackets (you can make them or buy them), so you need to make a blank plate to cover the hole in the chamber that's not being used. You can use a piece of ply about 100mm bigger all round than the hole. Paint one side of it with bituminous paint and the other with anything you like. When the bituminous paint has dried, fix a length of draughtproof strip around the edge and place it on the platform over the hole. You can fix it to the platform with some sort of brackets or just put something heavy on top.

On the next page are the instructions I suggest you display in your compost toilet to make sure it is used properly.

instructions

- **don't use just for a wee**
  Compost loos don't like too much liquid, so if all you want is a wee – please use one of the other toilets

- **please throw in a handful of sawdust after use**
  This balances the carbon to nitrogen ratio and helps bacteria to break the waste down

- **toilet paper is fine**
  And so is the middle of the toilet roll

a compost toilet is good for the environment in three main ways:

- it diverts human waste from sewage works and rivers – so it doesn't need to be treated with chlorine or other chemicals

- it allows organic matter to go back to the soil where it belongs

- it saves water – you're not flushing away one resource (fertilizer) with another (pure drinking water)

decorate

It's important to decorate really well, as people who may be put off by the thought of a compost loo will be even more put off if it's in a dirty or dingy room. Paint the room a bright, friendly colour (eco-paints of course), not forgetting the pipe lagging and vent pipe.

Although you don't need a toilet brush for the toilet itself, you will still need one for the urinal for when it is cleaned. Install a toilet roll holder, mirror, curtains, and maybe pictures, and a shelf with plants and pot pourri – anything to make it attractive so that people will want to use it. You should take more care than with a conventional toilet because visitors may have pre-conceived ideas and be reluctant to use a compost toilet. Make them feel more at ease by creating a welcoming, clean room. Imagine your granny using it.

extras

There are a couple of extras that you need in a compost toilet that you don't need in a conventional toilet. One is a bucket for sawdust / soak, and the other is instructions on the wall, preferably where they can be read whilst sitting on the loo, see model instructions on page 99, as people unused to compost loos may be looking for something to flush and may be unaware that they are supposed to throw down a handful of sawdust when they are finished.

Think about the position of lighting in the room. Wall lights may be a better idea than ceiling lights as, if the light is overhead, you may see just a little bit more than you want to when you look down the toilet.

Another good thing to have is a manual for your particular compost loo. This can be handed over to the new owners or tenants if the house is sold, or if someone looks after your house while you are travelling, for example. It should explain how to maintain and clean the toilet, where to get sawdust from, how to troubleshoot, peak-knock, empty the toilet, where to put the compost, how to change over the seat, the timescale for carrying out these tasks and anything else necessary to ensure that future users treat it in the way you intended it to be treated.

use and maintenance

- compost toilets require more maintenance than ordinary flush toilets; you have to do certain things and you have to think about certain things – it's not flush and forget.
- we want to spread the word about compost toilets and so they have to be pleasant to use. A smelly, dirty, uncomfortable compost toilet doesn't spread the word at all – quite the opposite. There shouldn't be anything nasty to see or any flies to escape when you open the lid.
- the more well-built and well-maintained compost toilets there are, the more people will come to accept them.
- people will not be familiar with them; gently introduce guests to the idea and make sure there are clear instructions on the wall.
- it may be a good idea to make some notes on the use and maintenance of your compost toilet, to be handed over to future owners, tenants or house-sitters.
- do remember to throw a handful of soak down after every use and don't throw anything else down, like sanitary towels or disposable nappies.

starting off
- put around 10cm of sawdust in the bottom of the chamber.
- some say it's a good idea to add a shovelful of soil. The millions of bacteria, fungi, moulds and other micro-organisms in soil will kick-start the composting process. I'm not sure about this one; the soil could contain the eggs of flies, fungus, gnats, mites, or other creatures you really don't want at this stage (it's fine when you seal the toilet after a year's use though). I have never added soil at the beginning and the composting process hasn't been hindered.

daily tasks
- check toilet paper etc. All the things you'd do with a conventional toilet, in other words.
- clean the room. This is very important, in fact more so than an ordinary loo, because some people will be a bit scared of it. It needs to be spotless to put people at their ease.
- clean round the seat and lid.
- check the sawdust in the bucket. Top up if necessary. If you have run out of sawdust, then get some more from a friendly sawmill or wood workshop. If you chainsaw firewood, then bag up the sawdust after every session and store it in a dry place.

weekly tasks

- empty the fly-trap if necessary. Unscrew the jar from the lid, and either vacuum up the flies (if there are any) or sweep them up with a brush and pan. You can tip dead flies into the toilet, but it's best not to do this if any are still alive.

monthly tasks

- check whether the 'peak' needs to be knocked. As you can imagine, because deposits are dropped in the same place a little 'peak' will build up, which needs to be knocked flat every now and then. You can do this via the hatch or, more easily, via the seat. Keep a special tool separate from the rest, preferably in a locked shed. A hoe is good. Put the hoe down into the chamber and move it from side to side, making sure that you flatten out the pile. Then take it outside and hose it down really well before locking it away again.
- check the insect screens on the end of the drain and the top of the vent pipe.
- check the ring of draughtproof strip on the toilet lid.

annual tasks

- empty the chamber that has been resting and decomposing for a year. Take off the hatch, get a long-handled shovel inside and empty the contents into a wheelbarrow. Don't worry, it will be just like ordinary compost, and there will be much less of it than you expect. This job is best done in winter – the heap will spend the summer breaking down and, if you put it around your fruit trees, flower beds etc, by the time the summer comes around again, it will be absorbed into the soil. If you are emptying the chamber in the summer, when there might be kids playing around your fruit trees etc, it may be best to bury the compost in a trench as there is a slight chance that there may be some pathogens present. See *pathogens / hygiene,* page 22 to get an idea of just how slim this chance is, but better safe than sorry.

fig 47: emptying the 'rested' chamber with a long-handled shovel into a wheelbarrow, for use on the Redfield apple trees

- you can leave it for more than a year if the other chamber isn't looking full – the longer the waste has to decompose the better. If you can get into a two-year cycle instead of a one-year cycle, then you're more certain to end up with good crumbly compost at the end.
- check that the drain is clear. You can clear it with a small drain rod if not.
- take the seat off the chamber you've been using and put it onto the now empty chamber. Fix the blank plate over the first chamber and leave it for a year to decompose.

fig 48: you should have thousands of compost worms like these in any compost heap; put a shovelful down the compost toilet when you've finished using it

- before you fix the blank plate, you could put a shovelful of worms into the full chamber. You can't do this before you've finished using it, as the worms don't like the salts in urine. But when you've finished using it it's OK. Don't use worms from your garden – composting worms are different from earthworms and are not found in soil – you can get them from your compost heap, or even from your septic tank if you have one. There are special composting worms that are smaller than earthworms, reddish in colour and with different shaded bands. There are brandling worms, tiger worms and – best of all – dendra worms (because they like to eat sawdust). If you can't find any in your compost heap, you can order some online see *resources*, page 113 – they come through the post!

- the worms will not only munch their way through the waste (breaking it down into smaller particles for micro-organisms to get at more easily), their digestive system will break down material to provide nutrients for plants, and they make tunnels which will allow air into the pile and provide oxygen for the aerobic micro-organisms. You can experiment with worms – put some in when you start, then see if they are still there when you empty the finished compost.

questions and troubleshooting

kitchen waste

Should I put kitchen / vegetable waste down the compost loo?
No. Use a compost heap instead. It seems silly to mix material that definitely has no pathogens with material that might. Vegetable waste can go on the compost heap and be used in the vegetable garden on any crops. Although the risk of pathogens remaining in finished compost from a compost toilet is tiny, see *pathogens / hygiene*, page 22, I still recommend using it for soft fruit, fruit trees and flower beds rather than the vegetable garden.

Also, vegetable scraps may have fly eggs on them and flies are more attracted to the smell of rotten vegetables than faeces, believe it or not (see below).

antibiotics

Will people on antibiotics using the compost toilet affect the ability of bacteria to decompose the waste material?
No. A study by the Department of Microbiology at the Agricultural School of Norway found that only a tiny percentage of antibiotics taken by a person will be excreted, and the dilution effect of the materials in the toilet will mean that the beneficial bacteria will hardly be affected at all.

smells

If your compost toilet smells unduly, then there is something wrong. If you have a vent, urine separation, a drain plus a sealed lid which is kept closed, and you add correct amounts of soak, then there shouldn't be any smells. If you have done all these things, and there are still smells, you can try the following:
- check the drain; you can use a little drain rod from outside, and when you empty the chamber you can do it from the inside too.
- maybe the vent isn't working properly: it shouldn't need a fan, but if you've tried everything else and there's still a smell it may be best to install one.

- check the pile isn't becoming too wet. Make sure people aren't weeing in it, and get them to add two handfuls of soak instead of one for a while to see if that solves the problem.
- make sure the lid is sealed properly and that it is always closed after use see *step-by-step DIY guide*, page 63.
- put a bowl of pot pourri in the room and change it regularly.
- put baking soda in the chamber.

red spider mites

These were a problem at Redfield for a while. They are tiny, barely visible, little red moving dots; they eat sawdust and almost definitely come in on damp sawdust. Make sure that the sawdust you bring in is dry.

We found that the only thing that killed them was the spray that we use to get rid of mites from our chickens. It is called red mite concentrate, and might be available at your local agricultural suppliers. If not, it is made by Barrier Biotech Ltd., see *resources*, page 113.

flies

Flies won't be attracted if you do everything right, as there won't be a smell. Also, you have to make sure that the lid is always closed when not in use and then flies won't be able to get in. To be on the safe side, you can put some mesh (nylon net curtain will do, fixed with electrical tape) over the end of the drain, and the vent (although even if you don't, spiders will make webs in the vent and drain pipe and catch them anyway).

Don't put vegetable scraps in toilet because:
- they may have flies or eggs on them
- flies are attracted to the smell of rotting vegetables more than they are attracted to faeces
If you do get flies:
- install a fly-catcher if you haven't already installed one
- use pyrethrum, derris or other organic insecticides
- as a very last resort, use fly-spray. I know it contains nasty chemicals, but if it means that otherwise you can't use the compost loo, then it just needs to be done. You shouldn't get to this point if you don't let them get in, but sometimes they manage to sneak in. Once you have them they can be difficult to get rid of. When they got into the Redfield wheelie bin loo we tried derris, an insecticide block and fly-paper, but it didn't get rid

of them all. So, as a last resort, we bought some fly-spray, opened the lid and gave the chamber a couple of sprays. It killed all the flies and, what's more, they didn't come back. We still have the fly-spray a couple of years later and haven't needed to use it again. If you use a spray, the amount involved will be so small relative to the size of the compost pile, that it won't affect the composting process at all.

the future

For poorer countries to adopt western sewerage systems, it would involve the use of technology, raw materials, energy, water and an enormous amount of money that they just don't have. As around a third of deaths in developing countries are attributable to poor sanitation and water-borne diseases, something needs to be done.

Compost toilets are almost definitely the best option – cheap, easily constructed and maintained, with low resource use and producing useful fertilizer. One possible barrier to their widespread use however, is the feeling amongst developing countries' populations and governments that compost toilets are somehow backward, and that flush systems seem to be alright for rich countries but not for them. But Michael Rouse, formerly the UK's chief drinking water inspector, recently said that if Britain were planning sewage disposal from scratch today, 'we wouldn't flush it away - we would collect the solids and compost it' (quoted in *New Scientist*).

This brings us to one of LILI's main principles – developed countries have to change first. They have the resources to test low-impact systems, and if they are accepted in richer countries, they will be more readily accepted elsewhere. We in the West have for too long been a model for developing countries of socially- and environmentally-damaging ways of living.

Although proprietary compost toilets are expensive for widespread use in developing countries, it's certainly a much cheaper option than a Western-style sewerage system, and they could initiate a self-build programme of twin-chamber compost loos.

As for the West – well there are some enlightened local authorities, and naturally they are in Scandinavia; the Swedish municipality of Tanum has decided that henceforth it will only give planning permission for compost toilets and not for conventional WCs.

There is a big opportunity for compost toilets in new-build housing. They can be quite difficult to retro-fit into existing houses and they won't get used much if outdoors. I firmly believe though that, one day, all new houses, instead of having a conventional sewer connection, will have a solid, reliable proprietary compost toilet (like a Clivus Multrum for example) as standard, as well as rainwater harvesting, greywater recycling (not to mention solar hot water, natural, local materials used for building, photovoltaics, wind turbine – I could go on and on). The hatch to the chamber could be on the outside of the building, and be emptied annually by a local authority truck, which would then deposit the compost on agricultural land. Homeowners could be charged a fee for emptying, unless they wanted to use the compost themselves on their garden.

For now, the large proprietary models are expensive. Local authorities could introduce subsidies or grants, but until they do, for real low-impacters, this book will help you to do it yourself.

resources

suppliers and manufacturers

compost toilets

Aquatron

www.aquatron.se
Swedish manufacturer of the Aquatron separator

Biolet

www.biolet.com
US manufacturer of the Biolet toilet

Clivus

www.clivus.com
Swedish manufacturer of the clivus multrum toilet

Eastwood Services

Kitty Mill
Wash Lane
Wenhaston
Halesworth
Suffolk IP19 9DX
www.sun-mar.com
01502 478165
British distributor of Sun-mar toilets

Elemental Solutions

www.elementalsolutions.co.uk
01981 540728
Aquatron, Dowmus, plus their own design toilet

Envirolet

www.envirolet.com
US manufacturer of small compost toilet

Kazuba

www.kazuba.eu
distributor of the award-winning enviroloo composting toilet unit

Kernowrat

www.kernowrat.co.uk
for the Biobag and Separett systems

Kingsley Clivus

Western Barn Industrial Park
Hatherleigh Rd
Winkleigh
Devon EX19 8AP
www.kingsleyplastics.co.uk/clivus%20home.htm
01837 83154
British distributor of Clivus Multrum

Maurice Moore Consultants

The Coach House
141 Hersham Road
Walton on Thames
Surrey KT12 1RW
01932 230763 / 0208 398 7951
British agent for Soltran and Rota-loo

NatSol Ltd

20 Bethel Street
Llanidloes
Powys SY18 6BS
www.natsol.co.uk
01686 412653
Manufacture and install their own twin-vault compost toilets

Nature's Head

www.natureshead.net
US manufacturer of small compost toilet

Rota-loo

www.rotaloo.com
Australian manufacturer of quite large toilet with rotating chambers

Sun-mar

www.sun-mar.com
US manufacturers of toilet with a heating element

Thunderboxes

www.thunderboxes2go.co.uk
07949 106332
Compost toilet hire for festivals

Wendage Pollution Control

Rangeways Farm
Conford, Liphook
Hampshire GU30 7QP
www.wpc.uk.net
01428 751296
British distributor of Biolet toilets

waterless urinals

Water Solutions Ltd

High Street
Sidcup, Kent
www.watersolution.com
0208 309 5556
They say that the valve that comes with their urinals needs changing every three months. This may be because their system is designed for offices, public buildings etc. and so has heavy usage. The Hepworth valves used at Redfield last several years.

Air Flush

www.greenbuildingstore.co.uk
01484 854898
Made by Solution Elements (the trading arm of 'Elemental Solutions', above), and supplied by the Green Building Store

materials for self-build

Salvo

www.salvo.co.uk
A list of salvage yards in Britain and internationally; salvage yards are excellent sources of recycled building materials. This site will help you find your local yard.

lime putty

You can purchase tubs of lime putty directly from LILI

worms

You can buy worms mail order, weird though worms-through-the-post sounds. If you don't have a compost heap to get worms from (to add to your compost loo chamber when you put it to 'rest') then you can order them for about £10 for 500 worms, which is more than enough. You can also get a wormery, which can be used to compost your kitchen waste even if you live in a flat, and you can take worms from that.

Vermisell

www.vermisell.co.uk

Wiggly Wigglers

www.wigglywigglers.co.uk

Worm City

www.wormcity.co.uk

Wormery Store
http://wormerystore.co.uk

red mite control

Red mite concentrate, from:

Barrier Hygiene Ltd

36 Haverscroft Industrial Estate
New Rd, Attleborough,
Norfolk, NR17 1YE
www.barrier-biotech.com
01953 456363

other

Joe's Plumbing
http://www.joes-plumbing.com/toilet_seat.html
US website selling 'self-closing toilet seats'

Watercourse Systems Ltd
Will's Barn
Chipstable
Taunton
Somerset TA4 2PX
01984 629 070

information

Adam Hart-Davis (the bloke on the telly) discusses loos,
and especially Henry Moule's earth closet
http://www.adam-hart-davis.org/loos.htm

Dowmus forum

http://dowmus.hinternet.com.au
A forum for dowmus owners having trouble with their unit

Environmental Design and Construction

http://www.edcmag.com/articles/
composting-toilets-emerge-as-viable-alternatives
Some interesting case studies

Humanure Headquarters

http://www.humanurehandbook.com
Information on basic compost toilets

Practical Action

www.itdg.org/docs/technical_information_service/compost_toilets.pdf
technical brief for the construction of compost toilets in developing
countries, from Practical Action - formerly the Intermediate Technology
Development Group

New Scientist

http://www.newscientist.com/article/dn3512-composting-toilets-key-to-global-sanitation-say-scientists.html
An article promoting compost loos as the best solution to developing countries' sewage problems and arguing that flush sewerage systems would be an environmental disaster. N.B. there are several articles on compost loos on the New Scientist website – just use the search facility.

Oasis Design

www.oasisdesign.net
Information on lots of environmental topics; there is a huge list of compost toilet suppliers around the world on the links page

Water Regulations Advisory Scheme

www.wras.co.uk
A website providing details of water regulations for the UK which includes an order form for the complete regulations that costs £16.30 plus postage and packing

World of Composting Toilets

http://compostingtoilet.org
This site contains useful information but only features Envirolet units

World of Composting Toilets UK

http://compostingtoilet.org.uk

This site contains information and links to purchase 3 types of toilet

Yahoo Group

http://groups.yahoo.com/group/compost-toilet/
online compost toilet forum

courses

LILI

www.lowimpact.org/compost_toilets
01296 714184
LILI has a range of compost toilet and sustainable water and sewage courses all over the UK that can be booked via their website

LILI also runs DIY for Beginners and Introduction to Plumbing courses

books

These books are available from LILI's online bookshop,
as long as they are in print
www.lowimpact.org/manuals.htm
LILI also sells books on most other environmental issues.

Lifting the Lid
Peter Harper and Louise Halestrap, CAT Publications, 1999

Fertile Waste
Peter Harper, CAT Publications, 1998

Sewage Solutions
Grant, Moodie and Weedon, CAT Publications, 1995

The Humanure Handbook
J C Jenkins, Jenkins Publishing, 1994
Irreverent look at turning human waste into useful compost

The Toilet Papers
Sim Van der Ryn, Ecological Design Press, 1978

The Composting Toilet System Book
David Del Porto and Carol Steinfeld,
The Center for Ecological Pollution Prevention, 2000,
Covers many different types of toilet that you can buy, and lots of case studies but uses a much more high-tech approach than this book

Sanitation Without Water
Uno Winblad and Wen Kilama, Macmillan, 1985
All about maintaining a septic tank / leachfield system, with chapters on alternatives, including compost toilets

Plumbing, Heating and Gas Installations
R D Treloar, Blackwell, 2000
The best plumbing book there is

regulations

Department of Environment

http://www.doeni.gov.uk/niea/ppg04.pdf
Pollution Prevention Guidance notes – PPG04 covers septic tanks, soakaways, leachfields, reed beds and compost toilets; N.B. government departments and agencies change their names and their websites every five minutes and this link is current at the time of publication but if it goes nowhere when you try to use it, just search for PPG04

other LILI publications

Learn how to heat your space and water using a renewable, carbon-neutral resource – wood.

This book includes everything you need to know, from planning your system, choosing, sizing, installing and making a stove, chainsaw use, basic forestry, health and safety, chimneys, pellet and woodchip stoves. The second edition has been expanded to reflect improvements in wood-fuelled appliances and the author's own recent experience of installing and using an automatic biomass system

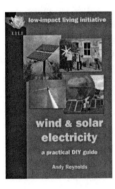

The author has been providing his own electricity from the sun and the wind for many years, and in this book he explains how you can do the same.

There are chapters on the various system components required (including inverters and charge controllers), how to put them all together, batteries, grid-connected systems, and there is even a basic electricity primer. Andy has analysed the output of his system for over 10 years, and these real-life figures are included.

The author grew up in Jamaica and was taught to make soaps by her grandmother. They grew all the plants they needed to scent and colour their soaps and even used wood ash from the stove to make caustic potash.

Her book is intended for beginners, includes both hot- and cold-process soap making, with careful step-by-step instructions, extensive bar, liquid and cream soap recipes, full details of equipment, a rebatching chapter plus information on the legislation and regulations for selling soap.

Now also available as part of an an online course at http://lowimpact.org/online_courses_natural_soaps.html

how to spin: just about anything is a wide-ranging introduction to an ancient craft which has very contemporary applications.
It tells you all you need to know about the available tools, from hand spindles to spinning wheels, what to do to start spinning, with illustrated, step-by-step instructions, and a comprehensive guide to the many fibres you can use to make beautiful yarns.

Janet Renouf-Miller is a registered teacher with the Association of Weavers, Spinners and Dyers, and has taught at their renowned Summer School.

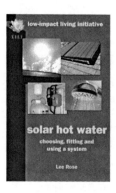

solar hot water: choosing, fitting and using a system provides detailed information about solar-heated water systems and is particularly applicable to domestic dwellings in the UK.
Lee Rose has 10 years of experience and involvement in every aspect of the solar thermal industry in the UK and around the world. His book provides a comprehensive introduction to every aspect of solar hot water: including all relevant equipment, components, system design and installation and even how to build your own solar panels.

herbal remedies: how to make, use and grow them teaches you to identify, grow and harvest medicinal plants.
It shows you how to make a range of simple medicines; there are sections on body systems, explaining which herbs are useful for a range of ailments, and detailed herb monographs. This second edition has been revised to take account of recent changes in UK legislation.

Sorrell Robbins is a highly-qualified, leading expert in natural health with over 15 years experience. She teaches at all levels from beginner to advanced postgraduate and is a regular contributor to many natural health publications.

CPSIA information can be obtained
at www.ICGtesting.com
Printed in the USA
FSOW02n0714280316
18523FS